PILOT ANALYSIS OF GLOBAL ECOSYSTEMS

# Agroecosystems

STANLEY WOOD

KATE SEBASTIAN

SARA J. SCHERR

**CAROL ROSEN**

*PUBLICATIONS DIRECTOR*


**HYACINTH BILLINGS**

*PRODUCTION MANAGER*


**MAGGIE POWELL**

*COVER DESIGN AND LAYOUT*


**CAROLLYNE HUTTER**

*EDITING*

**Photo Credits: Cover:** Viet Nam, IFPRI Photo/P. Berry, **Smaller ecosystem photos:** *Forests:* Digital Vision, Ltd., *Grasslands:* PhotoDisc, *Freshwater:* Dennis A. Wentz, *Coastal:* Digital Vision, Ltd., **Prologue:** Viet Nam, FAO Photo/P. Johnson, **Agricultural Extent and Agricultural Land Use Changes:** Rwanda, FAO Photo/A. Odoul, **Food, Feed, and Fiber:** Ecuador, FAO Photo/J. Bravo, **Soil Resource Condition:** Honduras, FAO Photo/G. Bizzarri, **Water Services:** China, FAO Photo/F. Botts, **Biodiversity:** Thailand, FAO Photo/A. Wolstad, **Carbon Services:** CGIAR Photo.

Pilot Analysis of Global Ecosystems

# Agroecosystems

Stanley Wood

Kate Sebastian

Sara J. Scherr

## With analytical contributions from:

Niels H. Batjes, ISRIC
Andrew Farrow, CIAT
Jean Marc Faurès, FAO
Günther Fischer IIASA
Gerhard Heilig, IIASA
Julio Henao, IFDC
Robert Hijmans, CIP
Freddy Nachtergaele, FAO
Peter Oram, IFPRI
Manuel Winograd, CIAT

A joint study by International Food Policy Research Institute and World Resources Institute

Published by International Food Policy Research Institute and World Resources Institute
Washington, DC

This report is also available at http://www.wri.org/wr2000 and http://www.ifpri.org

# Pilot Analysis of Global Ecosystems (PAGE)

*Project Management*
Norbert Henninger, WRI
Walt Reid, WRI
Dan Tunstall, WRI
Valerie Thompson, WRI
Arwen Gloege, WRI
Elsie Velez-Whited, WRI

*Agroecosystems*
Stanley Wood, International Food
  Policy Research Institute
Kate Sebastian, International Food
  Policy Research Institute
Sara J. Scherr, University of
  Maryland

*Coastal Ecosystems*
Lauretta Burke, WRI
Yumiko Kura, WRI
Ken Kassem, WRI
Mark Spalding, UNEP-WCMC
Carmen Revenga, WRI
Don McAllister, Ocean Voice
  International

*Forest Ecosystems*
Emily Matthews, WRI
Richard Payne, WRI
Mark Rohweder, WRI
Siobhan Murray, WRI

*Freshwater Systems*
Carmen Revenga, WRI
Jake Brunner, WRI
Norbert Henninger, WRI
Ken Kassem, WRI
Richard Payne, WRI

*Grassland Ecosystems*
Robin White, WRI
Siobhan Murray, WRI
Mark Rohweder, WRI

A series of five technical reports, available in print and on-line at http://www.wri.org/wr2000.

## AGROECOSYSTEMS

Stanley Wood, Kate Sebastian, and Sara J. Scherr, *Pilot Analysis of Global Ecosystems: Agroecosystems, A joint study by International Food Policy Research Institute and World Resources Institute*, International Food Policy Research Institute and World Resources Institute, Washington D.C.
December 2000 / paperback / ISBN 1-56973-457-7 / US$20.00

## COASTAL ECOSYSTEMS

Lauretta Burke, Yumiko Kura, Ken Kassem, Mark Spalding, Carmen Revenga, and Don McAllister, *Pilot Analysis of Global Ecosystems: Coastal Ecosystems*, World Resources Institute, Washington D.C.
November 2000 / paperback / ISBN 1-56973-458-5 / US$20.00

## FOREST ECOSYSTEMS

Emily Matthews, Richard Payne, Mark Rohweder, and Siobhan Murray, *Pilot Analysis of Global Ecosystems: Forest Ecosystems*, World Resources Institute, Washington D.C.
October 2000 / paperback / ISBN 1-56973-459-3 / US$20.00

## FRESHWATER SYSTEMS

Carmen Revenga, Jake Brunner, Norbert Henninger, Ken Kassem, and Richard Payne *Pilot Analysis of Global Ecosystems: Freshwater Systems*, World Resources Institute, Washington D.C.
October 2000 / paperback / ISBN 1-56973-460-7 / US$20.00

## GRASSLAND ECOSYSTEMS

Robin White, Siobhan Murray, and Mark Rohweder, *Pilot Analysis of Global Ecosystems: Grassland Ecosystems*, World Resources Institute, Washington D.C.
November 2000 / paperback / ISBN 1-56973-461-5 / US$20.00

The full text of each report will be available on-line at the time of publication. Printed copies may be ordered by mail from WRI Publications, P.O. Box 4852, Hampden Station, Baltimore, MD 21211, USA. To order by phone, call 1-800-822-0504 (within the United States) or 410-516-6963 or by fax 410-516-6998. Orders may also be placed on-line at http://www.wristore.com.

The agroecosystem report is also available at http://www.ifpri.org. Printed copies may be ordered by mail from the International Food Policy Research Institute, Communications Service, 2033 K Street, NW, Washington, D.C. 20006-5670, USA.

# Contents

MAPS

# TABLES

# FIGURES

# BOXES

# Foreword

Earth's ecosystems and its peoples are bound together in a grand and complex symbiosis. We depend on ecosystems to sustain us, but the continued health of ecosystems depends, in turn, on our use and care. Ecosystems are the productive engines of the planet, providing us with everything from the water we drink to the food we eat and the fiber we use for clothing, paper, or lumber. Yet, nearly every measure we use to assess the health of ecosystems tells us we are drawing on them more than ever and degrading them, in some cases at an accelerating pace.

Our knowledge of ecosystems has increased dramatically in recent decades, but it has not kept pace with our ability to alter them. Economic development and human well-being will depend in large part on our ability to manage ecosystems more sustainably. We must learn to evaluate our decisions on land and resource use in terms of how they affect the capacity of ecosystems to sustain life — not only human life, but also the health and productive potential of plants, animals, and natural systems.

A critical step in improving the way we manage the earth's ecosystems is to take stock of their extent, their condition, and their capacity to provide the goods and services we will need in years to come. To date, no such comprehensive assessment of the state of the world's ecosystems has been undertaken.

The Pilot Analysis of Global Ecosystems (PAGE) begins to address this gap. This study is the result of a remarkable collaborative effort between the World Resources Institute (WRI), the International Food Policy Research Institute (IFPRI), intergovernmental organizations, agencies, research institutes, and individual experts in more than 25 countries worldwide. The PAGE compares information already available on a global scale about the condition of five major classes of ecosystems: agroecosystems, coastal areas, forests, freshwater systems, and grasslands. IFPRI led the agroecosystem analysis, while the others were led by WRI. The pilot analysis examines not only the quantity and quality of outputs but also the biological basis for production, including soil and water condition, biodiversity, and changes in land use over time. Rather than looking just at marketed products, such as food and timber, the study also analyzes the condition of a broad array of ecosystem goods and services that people need, or enjoy, but do not buy in the marketplace.

The five PAGE reports show that human action has profoundly changed the extent, condition, and capacity of all major ecosystem types. Agriculture has expanded at the expense of grasslands and forests, engineering projects have altered the hydrological regime of most of the world's major rivers, settlement and other forms of development have converted habitats around the world's coastlines. Human activities have adversely altered the earth's most important biogeochemical cycles — the water, carbon, and nitrogen cycles — on which all life forms depend. Intensive management regimes and infrastructure development have contributed positively to providing some goods and services, such as food and fiber from forest plantations. They have also led to habitat fragmentation, pollution, and increased ecosystem vulnerability to pest attack, fires, and invasion by nonnative species. Information is often incomplete and the picture confused, but there are many signs that the overall capacity of ecosystems to continue to produce many of the goods and services on which we depend is declining.

The results of the PAGE are summarized in *World Resources 2000–2001*, a biennial report on the global environment published by the World Resources Institute in partnership with the United Nations Development Programme, the United Nations Environment Programme, and the World Bank. These institutions have affirmed their commitment to making the viability of the world's ecosystems a critical development priority for the 21st century. WRI and its partners began work with a conviction that the challenge of managing earth's ecosystems — and the consequences of failure — will increase significantly in coming decades. We end with a keen awareness that the scientific knowledge and political will required to meet this challenge are often lacking today. To make sound ecosystem management decisions in the future, significant changes are needed in the way we use the knowledge and experience at hand, as well as the range of information brought to bear on resource management decisions.

A truly comprehensive and integrated assessment of global ecosystems that goes well beyond our pilot analysis is necessary to meet information needs and to catalyze regional and local assessments. Planning for such a Millennium Ecosystem Assessment is already under way. In 1998, representatives from international scientific and political bodies began to explore the merits of, and recommend the structure for, such an assessment. After consulting for a year and considering the preliminary findings of the PAGE report, they concluded that an international scientific assessment of the present and likely future condition of the world's ecosystems was both feasible and urgently needed. They urged local, national, and international institutions to support the effort as stakeholders, users, and sources of expertise. If concluded successfully, the Millennium Ecosystem Assessment will generate new information, integrate current knowledge, develop methodological tools, and increase public understanding.

Human dominance of the earth's productive systems gives us enormous responsibilities, but great opportunities as well. The challenge for the 21st century is to understand the vulnerabilities and resilience of ecosystems, so that we can find ways to reconcile the demands of human development with the tolerances of nature.

We deeply appreciate support for this project from the Australian Centre for International Agricultural Research, The David and Lucile Packard Foundation, The Netherlands Ministry of Foreign Affairs, the Swedish International Development Cooperation Agency, the United Nations Development Programme, the United Nations Environment Programme, the Global Bureau of the United States Agency for International Development, and The World Bank.

A special thank you goes to the AVINA Foundation, the Global Environment Facility, and the United Nations Fund for International Partnerships for their early support of PAGE and the Millennium Ecosystem Assessment, which was instrumental in launching our efforts.

JONATHAN LASH
President
World Resources Institute

PER PINSTRUP-ANDERSEN
Director-General
International Food Policy Research Institute

# Acknowledgments

The World Resources Institute and the International Food Policy Research Institute would like to acknowledge the members of the Millennium Assessment Steering Committee, who generously gave their time, insights, and expert review comments in support of the Pilot Analysis of Global Ecosystems (PAGE):

Edward Ayensu, Ghana; Mark Collins, United Nations Environment Programme - World Conservation Monitoring Centre (UNEP-WCMC), United Kingdom; Angela Cropper, Trinidad and Tobago; Andrew Dearing, World Business Council for Sustainable Development (WBCSD); Janos Pasztor, UNFCCC; Louise Fresco, FAO; Madhav Gadgil, Indian Institute of Science, Bangalore, India; Habiba Gitay, Australian National University, Australia; Gisbert Glaser, UNESCO; Zuzana Guziova, Ministry of the Environment, Slovak Republic; He Changchui , FAO; Calestous Juma, Harvard University; John Krebs, National Environment Research Council, United Kingdom; Jonathan Lash, World Resources Institute; Roberto Lenton, UNDP; Jane Lubchenco, Oregon State University; Jeffrey McNeely, World Conservation Union (IUCN), Switzerland; Harold Mooney, International Council of Scientific Unions; Ndegwa Ndiangui, Convention to Combat Desertification; Prabhu L. Pingali, CIMMYT; Per Pinstrup-Andersen, Consultative Group on International Agricultural Research; Mario Ramos, Global Environment Facility; Peter Raven, Missouri Botanical Garden; Walter Reid, Secretariat; Cristian Samper, Instituto Alexander Von Humboldt, Colombia; José Sarukhán, CONABIO, Mexico; Peter Schei, Directorate for Nature Management, Norway; Klaus Töpfer, UNEP; José Galízia Tundisi, International Institute of Ecology, Brazil; Robert Watson, World Bank; Xu Guanhua, Ministry of Science and Technology, People's Republic of China; A.H. Zakri, Universiti Kebangsaan Malaysia, Malaysia.

The Pilot Analysis of Global Ecosystems would not have been possible without the data provided by numerous institutions and agencies. The authors of the agroecosystem analysis wish to express their gratitude for the generous cooperation and invaluable information we received from the following organizations:

Centro Internacional de Agricultura Tropical; Center for International Earth Science Information Network; Environmental Systems Research Institute; Food and Agriculture Organization of the United Nations; International Fertilizer Development Center; International Institute for Applied Systems Analysis; International Monetary Fund; International Potato Center; International Soil Reference and Information Centre; Technical Advisory Committee of the Consultative Group on International Agricultural Research; U.S. Department of Agriculture - NASS; United Nations Environment Programme; United Nations Environment Programme - World Conservation Monitoring Centre; U.S. Geological Survey (USGS) Earth Resources Observation System (EROS) Data Center; University of East Anglia; University of Kassel - Center for Environmental Systems Research; University of Maryland - Department of Geography; The World Bank; and World Wildlife Fund – U.S.A.

The authors would also like to express their gratitude to the many individuals who contributed information and advice, attended expert workshops, and reviewed successive drafts of this report.

Joseph Alcamo, University of Kassel; Carlos Baanante, International Fertilizer Development Center; K. Balasubramanian, JRD Tata Ecotechnology Centre; Niels Batjes, International Soil Reference and Information Centre; Christine Bergmark, United States Agency for International Development; Ruchi Bhandari, World Resources Institute; Ian Bowles, Conservation International; Jesslyn Brown, United States Geological Survey's EROS Data Center; Sally Bunning, Food and Agriculture Organization, AGLL; Emily Chalmers, consultant; Dan Claasen, United Nations Environment Programme; Linda Collette, Food and Agriculture Organization, SDRN; Braulio de Souza Dias, Ministry of the Environment, Brazil; Uwe Deichmann, The World Bank; John Dixon, The World Bank; Andrew Farrow, Centro Internacional de Agricultura Tropical; Jean Marc Faurès, Food and Agriculture Organization, AGLW; Günther Fischer, International Institute for Applied Systems Analysis; Louise Fresco, Food and Agriculture Organization, AGD; Robert Friedmann, The H. John Heinz III Center for Science, Economics, and the

Environment; Peter Frumhoff, Union of Concerned Scientists; Arthur Getz, World Resources Institute; Jack Gibbons; Luis Gomez, consultant; Richard Harwood, Michigan State University; Gerhard Heilig, International Institute for Applied Systems Analysis; Julio Henao, International Fertilizer Development Center; Robert Hijmans, International Potato Center; Tony Janetos, World Resources Institute; Peter Jones, Centro Internacional de Agricultura Tropical; Karen Jorgensen, United Nations Development Programme; Sjef Kauffman, International Soil Reference and Information Centre; Parviz Koohafkan, Food and Agriculture Organization, AGLL; Jonathan Loh, World Wide Fund for Nature International; Freddy Nachtergaele, Food and Agriculture Organization, AGLL; Robin O'Malley, The H. John Heinz III Center for Science, Economics, and the Environment; Stephen Prince, University of Maryland, Department of Geography; Don Pryor, Office of Science, Technology, and Policy; Armando Rabufetti, Inter-American Institute for Global Research; Mike Rodemeyr, Office of Science Technology and Policy; M.S. Swaminathan, M.S. Swaminathan Research Foundation; Ashbindu Singh, United Nations Environment Programme; Lori Ann Thrupp, Environmental Protection Agency; Thomas Walker, International Potato Center; Manuel Winograd, Centro Internacional de Agricultura Tropical; Hans Wolter, Food and Agriculture Organization, AGLD.

Not least, we would like to thank the many individuals at IFPRI and WRI who were generous with their time and comments as this report progressed: Connie Chan-Kang, Peter Hazell, Norbert Henninger, Emily Matthews, Siobhan Murray, Peter Oram, Phillip Pardey, Claudia Ringler, Mark Rosegrant, Melinda Smale, Kirsten Thompson and Liangzhi You. A special thank you goes to Mary-Jane Banks and Kathleen Flaherty for keeping us, and the project, in order throughout the report's production. A final thank you goes to Hyacinth Billings, Kathy Doucette, Carollyne Hutter, Michelle Morton and Maggie Powell, who provided editorial and design assistance, and guided the report through production.

# Introduction to the Pilot Analysis of Global Ecosystems

## PEOPLE AND ECOSYSTEMS

The world's economies are based on the goods and services derived from ecosystems. Human life itself depends on the continuing capacity of biological processes to provide their multitude of benefits. Yet, for too long in both rich and poor countries, development priorities have focused on how much humanity can take from ecosystems, and too little attention has been paid to the impact of our actions. We are now experiencing the effects of ecosystem decline in numerous ways: water shortages in the Punjab, India; soil erosion in Tuva, Russia; fish kills off the coast of North Carolina in the United States; landslides on the deforested slopes of Honduras; fires in the forests of Borneo and Sumatra in Indonesia. The poor, who often depend directly on ecosystems for their livelihoods, suffer most when ecosystems are degraded.

A critical step in managing our ecosystems is to take stock of their extent, their condition, and their capacity to continue to provide what we need. Although the information available today is more comprehensive than at any time previously, it does not provide a complete picture of the state of the world's ecosystems and falls far short of management and policy needs. Information is being collected in abundance but efforts are often poorly coordinated. Scales are noncomparable, baseline data are lacking, time series are incomplete, differing measures defy integration, and different information sources

may not know of each other's relevant findings.

## OBJECTIVES

The Pilot Analysis of Global Ecosystems (PAGE) is the first attempt to synthesize information from national, regional, and global assessments. Information sources include state of the environment reports; sectoral assessments of agriculture, forestry, biodiversity, water, and fisheries, as well as national and global assessments of ecosystem extent and change; scientific research articles; and various national and international data sets. The study reports on five major categories of ecosystems:

- Agroecosystems;
- Coastal ecosystems;
- Forest ecosystems;
- Freshwater systems;
- Grassland ecosystems.

These ecosystems account for about 90 percent of the earth's land surface, excluding Greenland and Antarctica. PAGE results are being published as a series of five technical reports, each covering one ecosystem. Electronic versions of the reports are posted on the Website of the World Resources Institute [http://www.wri.org/wr2000] and the agroecosystems report also is available on the Website of the International Food Policy Research Institute [http://www/ifpri.org].

The primary objective of the pilot analysis is to provide an overview of ecosystem condition at the global and continental levels. The analysis documents

the extent and distribution of the five major ecosystem types and identifies ecosystem change over time. It analyzes the quantity and quality of ecosystem goods and services and, where data exist, reviews trends relevant to the production of these goods and services over the past 30 to 40 years. Finally, PAGE attempts to assess the capacity of ecosystems to continue to provide goods and services, using measures of biological productivity, including soil and water conditions, biodiversity, and land use. Wherever possible, information is presented in the form of indicators and maps.

A second objective of PAGE is to identify the most serious information gaps that limit our current understanding of ecosystem condition. The information base necessary to assess ecosystem condition and productive capacity has not improved in recent years, and may even be shrinking as funding for environmental monitoring and record-keeping diminishes in some regions.

Most importantly, PAGE supports the launch of a Millennium Ecosystem Assessment, a more ambitious, detailed, and integrated assessment of global ecosystems that will provide a firmer basis for policy- and decision-making at the national and subnational scale.

## AN INTEGRATED APPROACH TO ASSESSING ECOSYSTEM GOODS AND SERVICES

Ecosystems provide humans with a wealth of goods and services, including food, building and clothing materials, medicines, climate regulation, water purification, nutrient cycling, recreation opportunities, and amenity value. At present, we tend to manage ecosystems for one dominant good or service, such as grain, fish, timber, or hydropower, without fully realizing the trade-offs we are making. In so doing, we may be sacrificing goods or services more valuable than those we receive — often those goods and services that are not yet valued in the market, such as biodiversity and flood control. An integrated ecosystem approach considers the entire range of possible goods and services a given ecosystem provides and attempts to optimize the benefits that society can derive from that ecosystem and across ecosystems. Its purpose is to help make trade-offs efficient, transparent, and sustainable.

Such an approach, however, presents significant methodological challenges. Unlike a living organism, which might be either healthy or unhealthy but cannot be both simultaneously, ecosystems can be in good condition for producing certain goods and services but in poor condition for others. PAGE attempts to evaluate the condition of ecosystems by assessing separately their capacity to provide a variety of goods and services and examining the trade-offs humans have made among those goods and services. As one example, analysis of a particular region might reveal that food production is high but, because of irrigation and heavy fertilizer application, the ability of the system to provide clean water has been diminished.

Given data inadequacies, this systematic approach was not always feasible. For each of the five ecosystems, PAGE researchers, therefore, focus on documenting the extent and distribution of ecosystems and changes over time. We develop indicators of ecosystem condition — indicators that inform us about

the current provision of goods and services and the likely capacity of the ecosystem to continue providing those goods and services. Goods and services are selected on the basis of their perceived importance to human development. Most of the ecosystem studies examine food production, water quality and quantity, biodiversity, and carbon sequestration. The analysis of forests also studies timber and woodfuel production; coastal and grassland studies examine recreational and tourism services; and the agroecosystem study reviews the soil resource as an indicator of both agricultural potential and its current condition.

## PARTNERS AND THE RESEARCH PROCESS

The Pilot Analysis of Global Ecosystems was a truly international collaborative effort. The World Resources Institute and the International Food Policy Research Institute carried out their research in partnership with numerous institutions worldwide (see *Acknowledgments*). In addition to these partnerships, PAGE researchers relied on a network of international experts for ideas, comments, and formal reviews. The research process included meetings in Washington, D.C., attended by more than 50 experts from developed and developing countries. The meetings proved invaluable in developing the conceptual approach and guiding the research program toward the most promising indicators given time, budget, and data constraints. Drafts of PAGE reports were sent to over 70 experts worldwide, presented and critiqued at a technical meeting of the Convention on Biological Diversity in Montreal (June, 1999) and discussed at a Millennium Assessment planning meeting in Kuala Lumpur, Malaysia (September, 1999). Draft PAGE materials and indicators were also presented

and discussed at a Millennium Assessment planning meeting in Winnipeg, Canada, (September, 1999) and at the meeting of the Parties to the Convention to Combat Desertification, held in Recife, Brazil (November, 1999).

## KEY FINDINGS
Key findings of PAGE relate both to ecosystem condition and the information base that supported our conclusions.

## The Current State of Ecosystems
The PAGE reports show that human action has profoundly changed the extent, distribution, and condition of all major ecosystem types. Agriculture has expanded at the expense of grasslands and forests, engineering projects have altered the hydrological regime of most of the world's major rivers, settlement and other forms of development have converted habitats around the world's coastlines.

The picture we get from PAGE results is complex. Ecosystems are in good condition for producing some goods and services but in poor condition for producing others. Overall, however, there are many signs that the capacity of ecosystems to continue to produce many of the goods and services on which we depend is declining. Human activities have significantly disturbed the global water, carbon, and nitrogen cycles on which all life depends. Agriculture, industry, and the spread of human settlements have permanently converted extensive areas of natural habitat and contributed to ecosystem degradation through fragmentation, pollution, and increased incidence of pest attacks, fires, and invasion by nonnative species.

The following paragraphs look across ecosystems to summarize trends in production of the most important goods and services and the outlook for ecosystem productivity in the future.

### Food Production

Food production has more than kept pace with global population growth. On average, food supplies are 24 percent higher per person than in 1961 and real prices are 40 percent lower. Production is likely to continue to rise as demand increases in the short to medium term. Long-term productivity, however, is threatened by increasing water scarcity and soil degradation, which is now severe enough to reduce yields on about 16 percent of agricultural land, especially cropland in Africa and Central America and pastures in Africa. Irrigated agriculture, an important component in the productivity gains of the Green Revolution, has contributed to waterlogging and salinization, as well as to the depletion and chemical contamination of surface and groundwater supplies. Widespread use of pesticides on crops has lead to the emergence of many pesticide-resistant pests and pathogens, and intensive livestock production has created problems of manure disposal and water pollution. Food production from marine fisheries has risen sixfold since 1950 but the rate of increase has slowed dramatically as fisheries have been overexploited. More than 70 percent of the world's fishery resources for which there is information are now fully fished or overfished (yields are static or declining). Coastal fisheries are under threat from pollution, development, and degradation of coral reef and mangrove habitats. Future increases in production are expected to come largely from aquaculture.

### Water Quantity

Dams, diversions, and other engineering works have transformed the quantity and location of freshwater available for human use and sustaining aquatic ecosystems. Water engineering has profoundly improved living standards, by providing fresh drinking water, water for irrigation, energy, transport, and flood control. In the twentieth century, water withdrawals have risen at more than double the rate of population increase and surface and groundwater sources in many parts of Asia, North Africa, and North America are being depleted. About 70 percent of water is used in irrigation systems where efficiency is often so low that, on average, less than half the water withdrawn reaches crops. On almost every continent, river modification has affected the flow of rivers to the point where some no longer reach the ocean during the dry season. Freshwater wetlands, which store water, reduce flooding, and provide specialized biodiversity habitat, have been reduced by as much as 50 percent worldwide. Currently, almost 40 percent of the world's population experience serious water shortages. Water scarcity is expected to grow dramatically in some regions as competition for water grows between agricultural, urban, and commercial sectors.

### Water Quality

Surface water quality has improved with respect to some pollutants in developed countries but water quality in developing countries, especially near urban and industrial areas, has worsened. Water is degraded directly by chemical or nutrient pollution, and indirectly when land use change increases soil erosion or reduces the capacity of ecosystems to filter water. Nutrient runoff from agriculture is a serious problem around the world, resulting in eutrophication and human health hazards in coastal regions, especially in the Mediterranean, Black Sea, and northwestern Gulf of Mexico. Water-borne diseases caused by fecal contamination of water by untreated sewage are a major source of morbidity and mortality in the developing world. Pollution and the introduction of non-native species to freshwater ecosystems have contributed to serious declines in freshwater biodiversity.

### Carbon Storage

The world's plants and soil organisms absorb carbon dioxide ($CO_2$) during photosynthesis and store it in their tissues, which helps to slow the accumulation of $CO_2$ in the atmosphere and mitigate climate change. Land use change that has increased production of food and other commodities has reduced the net capacity of ecosystems to sequester and store carbon. Carbon-rich grasslands and forests in the temperate zone have been extensively converted to cropland and pasture, which store less carbon per unit area of land. Deforestation is itself a significant source of carbon emissions, because carbon stored in plant tissue is released by burning and accelerated decomposition. Forests currently store about 40 percent of all the carbon held in terrestrial ecosystems. Forests in the northern hemisphere are slowly increasing their storage capacity as they regrow after historic clearance. This gain, however, is more than offset by deforestation in the tropics. Land use change accounts for about 20 percent of anthropogenic carbon emissions to the atmosphere. Globally, forests today are a net source of carbon.

### Biodiversity

Biodiversity provides many direct benefits to humans: genetic material for crop and livestock breeding, chemicals for medicines, and raw materials for industry. Diversity of living organisms and the abundance of populations of many species are also critical to maintaining biological services, such as pollination and nutrient cycling. Less tangibly, but no less importantly, diversity in nature is regarded by most people as valuable in its own right, a source of aesthetic pleasure, spiritual solace, beauty, and wonder. Alarming losses in global biodiversity have occurred over the past century. Most are the result of habitat destruction. Forests, grasslands, wetlands, and mangroves have been extensively converted to other uses; only tundra, the Poles, and deep-sea ecosystems have experienced relatively little change. Biodiversity has suffered as agricultural land, which supports far less biodiversity than natural forest, has expanded primarily at the expense of forest areas. Biodiversity is also diminished by intensification, which reduces the area allotted to hedgerows, copses, or wildlife corridors and displaces traditional varieties of seeds with modern high-yielding, but genetically uniform, crops. Pollution, overexploitation, and competition from invasive species represent further threats to biodiversity. Freshwater ecosystems appear to be the most severely degraded overall, with an estimated 20 percent of freshwater fish species becoming extinct, threatened, or endangered in recent decades.

## Information Status and Needs

### Ecosystem Extent and Land Use Characterization

Available data proved adequate to map approximate ecosystem extent for most regions and to estimate historic change in grassland and forest area by comparing current with potential vegetation cover. PAGE was able to report only on recent changes in ecosystem extent at the global level for forests and agricultural land.

PAGE provides an overview of human modifications to ecosystems through conversion, cultivation, firesetting, fragmentation by roads and dams, and trawling of continental shelves. The study develops a number

of indicators that quantify the degree of human modification but more information is needed to document adequately the nature and rate of human modifications to ecosystems. Relevant data at the global level are incomplete and some existing data sets are out of date.

Perhaps the most urgent need is for better information on the spatial distribution of ecosystems and land uses. Remote sensing has greatly enhanced our knowledge of the global extent of vegetation types. Satellite data can provide invaluable information on the spatial pattern and extent of ecosystems, on their physical structure and attributes, and on rates of change in the landscape. However, while gross spatial changes in vegetation extent can be monitored using coarse-resolution satellite data, quantifying land cover change at the national or subnational level requires high-resolution data with a resolution of tens of meters rather than kilometers.

Much of the information that would allow these needs to be met, at both the national and global levels, already exists, but is not yet in the public domain. New remote sensing techniques and improved capabilities to manage complex global data sets mean that a complete satellite-based global picture of the earth could now be made available, although at significant cost. This information would need to be supplemented by extensive ground-truthing, involving additional costs. If sufficient resources were committed, fundamentally important information on ecosystem extent, land cover, and land use patterns around the world could be provided at the level of detail needed for national planning. Such information would also prove invaluable to international environmental conventions, such as those dealing with wetlands, biological diversity, desertification, and climate change, as well as the international agriculture, forest, and fishery research community.

### Ecosystem Condition and Capacity to Provide Goods and Services

In contrast to information on spatial extent, data that can be used to analyze ecosystem condition are often unavailable or incomplete. Indicator development is also beset by methodological difficulties. Traditional indicators, for example, those relating to pressures on environments, environmental status, or societal responses (pressure-state-response model indicators) provide only a partial view and reveal little about the underlying capacity of the ecosystem to deliver desired goods and services. Equally, indicators of human modification tell us about changes in land use or biological parameters, but do not necessarily inform us about potentially positive or negative outcomes.

Ecosystem conditions tend to be highly site-specific. Information on rates of soil erosion or species diversity in one area may have little relevance to an apparently similar system a few miles away. It is expensive and challenging to monitor and synthesize site-specific data and present it in a form suitable for national policy and resource management decisions. Finally, even where data are available, scientific understanding of how changes in biological systems will affect goods and services is limited. For example, experimental evidence shows that loss of biological diversity tends to reduce the resilience of a system to perturbations, such as storms, pest outbreaks, or climate change. But scientists are not yet able to quantify how much resilience is lost as a result of the loss of biodiversity in a particular site or how that loss of resilience might affect the long-term provision of goods and services.

Overall, the availability and quality of information tend to match the recognition accorded to various goods and services by markets. Generally good data are available for traded goods, such as grains, fish, meat, and timber products and some of the more basic relevant productivity factors, such as fertilizer application rates, water inputs, and yields. Data on products that are exchanged in informal markets, or consumed directly, are patchy and often modeled. Examples include fish landings from artisanal fisheries, woodfuels, subsistence food crops and livestock, and nonwood forest products. Information on the biological factors that support production of these goods — including size of fish spawning stocks, biomass densities, subsistence food yields, and forest food harvests — are generally absent.

The future capacity (long-term productivity) of ecosystems is influenced by biological processes, such as soil formation, nutrient cycling, pollination, and water purification and cycling. Few of these environmental services have, as yet, been accorded economic value that is recognized in any functioning market. There is a corresponding lack of support for data collection and monitoring. This is changing in the case of carbon storage and cycling. Interest in the possibilities of carbon trading mechanisms has stimulated research and generated much improved data on carbon stores in terrestrial ecosystems and the dimensions of the global carbon cycle. Few comparable data sets exist for elements such as nitrogen or sulfur, despite their

fundamental importance in maintaining living systems.

Although the economic value of genetic diversity is growing, information on biodiversity is uniformly poor. Baseline and trend data are largely lacking; only an estimated 15 to 20 percent of the world's species have been identified. The OECD Megascience Forum has launched a new international program to accelerate the identification and cataloging of species around the world. This information will need to be supplemented with improved data on species population trends and the numbers and abundance of invasive species. Developing databases on population trends (and threat status) is likely to be a major challenge, because most countries still need to establish basic monitoring programs.

The PAGE divides the world's ecosystems to examine them at a global scale and think in broad terms about the challenges of managing them sustainably. In reality, ecosystems are linked by countless flows of material and human actions. The PAGE analysis does not make a distinction between natural and managed ecosystems; human intervention affects all ecosystems to some degree. Our aim is to take a first step toward understanding the collective impacts of those interventions on the full range of goods and services that ecosystems provide. We conclude that we lack

much of the baseline information necessary to determine ecosystem conditions at a global, regional or, in many instances, even a local scale. We also lack systematic approaches necessary to integrate analyses undertaken at different locations and spatial scales.

Finally, it should be noted that PAGE looks at past trends and current status, but does not try to project future situations where, for example, technological development might increase dramatically the capacity of ecosystems to deliver the goods and services we need. Such considerations were beyond the scope of the study. However, technologies tend to be developed and applied in response to market-related opportunities. A significant challenge is to find those technologies, such as integrated pest management and zero tillage cultivation practices in the case of agriculture, that can simultaneously offer market-related as well as environmental benefits. It has to be recognized, nonetheless, that this type of "win-win" solution may not always be possible. In such cases, we need to understand the nature of the trade-offs we must make when choosing among different combinations of goods and services. At present our knowledge is often insufficient to tell us where and when those trade-offs are occurring and how we might minimize their effects.

# Agroecosystems: Executive Summary

## Scope of Assessment

This study analyzes quantitative and qualitative information and develops selected indicators of the condition of the world's agroecosystems. We assess condition in terms of the delivery of a number of key goods and services valued by society: food, feed and fiber; water services; biodiversity; and carbon storage. We also attempt to assess pressures on, and current state of the underlying natural resource base. To this end we include an additional section dealing with soil resource condition, both as a determinant of agroecosystem capacity to produce goods and services and as a consequence of agroecosystem management practices.

### AGROECOSYSTEM EXTENT AND CHANGE

We define an agroecosystem as *"a biological and natural resource system managed by humans for the primary purpose of producing food as well as other socially valuable nonfood goods and environmental services."* The study first locates the global extent of agricultural lands within which these ecosystems are situated, using a satellite-derived land cover database common to all PAGE ecosystem studies. Although this data source allows agricultural extent to be approximated, with some significant regional limitations, it is not adequate for systematically identifying major agricultural land cover types within the agricultural extent. Furthermore, data on actual production systems and management practices—critical to understanding the potential sustainability and environmental consequences of agriculture—do not exist at regional or global scales (and certainly not in spatial formats). Thus, our representation of agroecosystems is generally limited to tabular summaries at a commodity or major land use level of aggregation and in a spatial context that reflects geopolitical rather than production system sub-divisions.

Conversion of forest and grassland for agriculture has had significant impacts on all goods and services. The predominantly positive effects on food outputs have usually been matched by correspondingly negative effects on the provision of water, biodiversity, and carbon storage services and on the quality of the soil resource. We developed several indicators that reflect the status of these changes and that flag land-use trade-offs in pursuing extensive versus intensive agriculture. Increasing yield correspondingly reduces land conversion pressures, but implies the adoption of more intensive production practices that might lead to other negative environmental effects.

### AGROECOSYSTEM GOODS AND SERVICES

Indicators of condition and value for agricultural and environmental goods and services were determined in consultation with agroecosystem experts in consideration of data limitations. Wherever possible, the spatial distribution of indicator values is shown within the satellite-derived global extent of agriculture, and indicator trends are presented in tabular formats. Indicators are developed for the following:

♦ Agricultural extent and the climatic characterization of agricultural land,
♦ Food, feed, and fiber land use and yields,
♦ Intensity of production input use,
♦ Value of food outputs in monetary, nutrition, employment, and income terms,
♦ Soil resource quality,
♦ Irrigation water use and efficiency,
♦ Watershed modification,
♦ Habitat conversion,
♦ Habitat quality of agricultural land,
♦ Agrobiodiversity,
♦ Carbon storage in agricultural soils and vegetation, and
♦ Agriculture's role in greenhouse gas emissions.

Many issues are not represented or are poorly represented in the above list. These include water quality, threats to wild terrestrial and aquatic biodiversity, human health, and impacts on the global nitrogen cycle. Livestock and agroforestry-based agroecosystems are also poorly represented, largely because of the difficulties of adequately locating extensive pasture and tree crops with the available satellite database. As will become evident, better data exist for those factors related directly to food production and markets. Environmental and ecological issues *within agroecosystems* have, until relatively recently, received little attention and relevant data is scarce above the local level.

Some indicators are related to multiple goods and services. For example, pesticide use relates to enhanced crop yields, reduced biodiversity, and increased water pollution. An increase in fertilizer application rate from an existing low level may lead to greater food production, improved soil fertility directly and indirectly (by producing more crop residues), and pose no threat to water pollution. However, that same increase at an already high level of application may provide a limited gain in productivity and pose an aquatic pollution threat. Caution is required, therefore, in interpreting any change in indicator value as representing a better situation when it actually may reflect a worse situation from the perspective of another good or service. This underscores the likely existence of trade-offs when attempting to enhance multiple agroecosystem outputs. Furthermore, correlation between indicators can make it difficult to interpret simultaneous changes in multiple indicators. One way to limit this difficulty, not presented here, is to posit the likely direction and extent of change in the value of each indicator that might arise from a given change in each of the other indicators. This would greatly aid the interpretation of observed changes in multiple indicators.

Because different levels and mixes of management skill, scientific and local knowledge, and other inputs can produce a different range of goods and service from the same set of natural resource inputs, there is no single measure of agroecosystem capacity even for a single good or service. Capacity is continually being redefined by the adoption of new technologies and resource management strategies, as well as by evolving institutions and markets, and changing agricultural and environmental policies. One example presented in this report is the observed increase in overall agroecosystem capacity to produce cereals as measured by a yield indicator, with a simultaneous decline in the long-term biophysical capacity because of continuous nutrient mining. Each section concludes, therefore, with a brief qualitative review of the factors influencing capacity, but we do not attempt to quantify current or future capacity per se. The proposed Millennium Ecosystem Assessment will go beyond this PAGE study in part by developing ecosystem management scenarios that should allow one to evaluate agroecosystem capacity in specific contexts.

## Key Findings and Information Issues

The following tables (pp. 3–8) summarize key findings of the study regarding the condition of agroecosystems, as well as the quality and availability of data.

# Agroecosystems Extent and Change

| PAGE MEASURES AND INDICATORS | DATA SOURCES AND COMMENTS |
|---|---|
| Extent and Area Intensity of Agriculture | Reinterpretation of Global Land Cover Characterization Data - 1km resolution satellite data for the period April 1992 to March 1993 set (GLCCD 1998; USGS EDC 1999a). |
| Agricultural Land Use Balance and Trends | FAO national tabular data for 1965-97 (FAOSTAT 1999). |
| Agroclimatic Factors and Generalized Slope | Global Agroecological Zones (GAEZ) database at half degree resolution (FAO/IIASA 1999). Includes climate variables based on 30 years of monthly data (UEA 1998). |
| Percent and Area of Land Equipped for Irrigation | University of Kassel global spatial data at half degree resolution (Döll and Siebert 1999). |
| Generalized Agroecosystem Characterization | Combination of PAGE data on extent, agroclimate, slope, and irrigation. |

## CONDITIONS AND TRENDS

♦ Cropland and managed pasture detected by satellite interpretation cover some 28 percent of global land surface. Globally, land with greater than 60 percent agriculture occupies 41 percent of the PAGE agricultural extent, while land with 40–60 percent and 30–40 percent agriculture occupies 29 and 30 percent of the PAGE agricultural extent, respectively.

♦ Overall, 31 percent of agricultural areas are occupied by crops and the remaining 69 percent are under pasture. Annual cropland is relatively stable at around 1.38 billion hectares, while permanent crops occupy around 131 million hectares (Mha) and show a net growth of almost 2 percent per year. Pasture areas are estimated to be increasing at around 0.3 percent per year.

♦ 91 percent of cropland is under annual crops, such as wheat, while perennial crops, such as citrus and tea, occupy the remainder. Although annual cropland is stable, the harvested area of annual crops is increasing at around 0.3 percent per year. The cropping intensity for annual crops globally currently stands at around 0.8.

♦ Irrigated areas occupy 270 Mha, around 5.4 percent of global agricultural land and 17.5 percent of all cropland. Irrigated area continues to expand, but at a slowing rate, now around 1.6 percent (about 3.3 Mha) per year. This net amount is presumed to allow for irrigated area losses—estimated as up to 1.5 Mha per year from salinization.

♦ 38 percent of the area within the satellite derived global extent of agriculture is found in temperate regions, another 38 percent in tropical regions, and some 23 percent in subtropical regions.

## INFORMATION STATUS AND NEEDS

♦ The resolution and interpretation of available global satellite data is generally insufficient to reliably distinguish all types of agricultural land cover. Extensive pasture, irrigated areas, fallow lands, and farmed transition areas between cropland and forest are particularly problematic. Finer resolution satellite data, expected to be available over the coming years, should improve our ability to reliably detect major agricultural land cover types.

♦ The utility of remotely sensed data would improve with more frequent interpretations (e.g., every three to five years) which focus on detecting land cover change at a global scale. This would require improved data resolution, more systematic classification processes, and innovative approaches to ground-truthing.

♦ The consistency and reliability of FAO national (tabular) agricultural land use data vary significantly. FAO no longer reports pasture and forest land cover types. Furthermore, national level agricultural land use data provide insufficient spatial disaggregation to help characterize agroecosystems.

♦ Looking solely at year-to-year *net* changes in land use gives an unrealistically conservative impression of the true dynamics of land use change. Additional information is required on conversions to and from agricultural land as well as between important agricultural land uses.

♦ The University of Kassel has developed the best available digital map of irrigation at a global scale, but the accuracy of the map is variable because of inconsistent scale, age, and reliability of source data. Efforts are continuing to improve this data set (e.g., by linking to published and unpublished elements of FAO's AQUASTAT databases).

♦ Data on production systems and resource management aspects of land use are extremely scarce at regional and global scales. Proxy measures include the following: statistical data on crop types, areas, and yields; animal populations and products; and the use of labor, irrigation, fertilizers, pesticides, and modern crop varieties. Although such data provide broad notions of land use and management, they are often difficult to obtain at subnational levels, are seldom interrelated, and give no indication of the scale of enterprise, the temporal and spatial arrangements of production systems, conservation practices, and so on.

# Food, Feed, and Fiber

| PAGE MEASURES AND INDICATORS | DATA SOURCES AND COMMENTS |
|---|---|
| **Production and Productivity:** Crop and pasture areas *(hectares)* Yields *(metric tons per hectare)* | FAO national tabular data for 1965-97 (FAOSTAT 1999). |
| **Intensity of Input Use:** *(see Water Services for irrigation indicators)* Fertilizer application - Nitrogen (N), phosphorus ($P_2O_5$), and potassium ($K_2O$): NPK *(kg per hectare)* | FAO national tabular data for 1965-97 (FAOSTAT 1999). |
| Pesticide application *(kilograms per hectare)* | Yudelman 1998. |
| Labor *(agricultural workers per hectare)* | FAO national tabular data for 1965-97 (FAOSTAT 1999). |
| Tractors *(hectares per tractor)* | FAO national tabular data for 1965-97 (FAOSTAT 1999). |
| **Value of Agricultural Production (VoP):** Total VoP *($)* VoP per hectare of cropland *($ per hectare)* | FAO national tabular data for 1965-97 (FAOSTAT 1999); Prices: (FAO 1997a). |
| **Nutritional Value:** Calories, protein, and fat per person | FAO Food Balance Sheets (FAOSTAT 1999) and FAO World Food Survey (1996). |
| **Employment and Income:** Number of agricultural workers Value of production per agricultural worker | FAO national tabular data for 1965-97 (FAOSTAT 1999). World Development Indicators (World Bank 2000). |

## CONDITIONS AND TRENDS

♦ Food production from agroecosystems is valued at around $1.3 trillion per year (1997) and provides 94 percent of the protein and 99 percent of the calories consumed by humans. The production process directly employs some 1.3 billion people.

♦ Food production has more than kept pace with global population growth. On average, food supplies are 24 percent higher per person than in 1961, and real prices are 40 percent lower. Over the same period, the global population has doubled from 3 to 6 billion people. Approximately 790 million people in the developing world are still chronically undernourished, almost two-thirds of whom reside in Asia and the Pacific.

♦ Although the global expansion of agricultural area has been modest in recent decades, intensification has been rapid, as irrigated area increased, fallow time decreased, and the use of purchased inputs and new technologies grew to produce more output per hectare.

♦ The dominant share of the world's cropland (59 percent) is dedicated to cereal production, but cereal yield growth rates have generally slowed in recent times.

♦ Over the past 30 years, the quantity of livestock products has approximately tripled compared to a doubling of crop outputs. This high rate of growth in livestock demand is expected to continue as, globally, standards of living and average incomes continue to rise.

♦ Increased demand for both crop and livestock products will come predominantly from developing countries, and because of infrastructure, institutional, and other trade and marketing constraints, will often need to be met from improved local agroecosystem capacity.

♦ The generally positive current trends in food production may mask negative trends in the underlying biophysical capacity of agroecosystems, e.g., nutrient mining, soil erosion, and overextraction of groundwater resources. But natural resource data is often too limited in space and time to gauge the full scope of such impacts.

♦ Environmental problems often associated with high-input agroecosystems include salinization of irrigated areas, nutrient and pesticide leaching, and pesticide resistance. Those more associated with low-input and extensive agroecosystems include soil erosion and loss of soil fertility.

♦ The specific mix of inputs and production technology has a direct bearing on the long-term capacity of agroecosystems to provide goods and services. Management practices can change rapidly in response to market signals and new technological opportunities, and can compensate for some aspects of resource degradation. Resource degradation increases reliance on the use of external (often purchased) inputs to maintain production levels.

## INFORMATION STATUS AND NEEDS

♦ The FAO tabular data was taken from national statistics but consistency and reliability among countries and years may vary. Additionally, the geopolitical basis of area and yield data limit their use for agroecosystem assessments that require such data be reaggregated into agroecosystem units.

♦ Enormous regional disparities exist among the indicators of yield, nutrition, value, and income. More local data is essential to better appreciate the nature and source of these differences and to target appropriate food and environment related policy and technology interventions.

♦ Soil nutrient balances complement yield data to provide a better understanding of agroecosystem condition. Soil nutrient balance trends, characterized by production systems (not simply by commodity) and by nutrient source (both inorganic and organic), would greatly enrich debates on the design and targeting of appropriate, integrated nutrient strategies.

♦ Relative to the other goods and services considered in this study, data on food, feed, and fiber is the most complete.

# Soil Resource Condition

## PAGE MEASURES AND INDICATORS

## DATA SOURCES AND COMMENTS

**Inherent Soil Constraints:**
Dominant soil fertility constraints
Area free of soil constraints

Fertility Capability Classification (FCC) (Sanchez et al. 1982; Smith 1989; Smith et al. 1997). FCC modifiers assessed for each soil mapping unit (SMU) of FAO's Digital Soil Map of the World at a resolution of 5x5 km (FAO 1995). SMUs comprise multiple soil types whose area shares but not location are known. This limits the spatial interpretation of such data.

**Status and Change in Soil Quality:**
Severity (extent and degree) of soil degradation

Global and regional (South and Southeast Asia) assessments of human-induced soil degradation (GLASOD: Oldeman et al. 1991a; and ASSOD: Van Lynden and Oldeman 1997) at scales of 1:10m and 1:5m respectively. Data based mainly on expert opinion for interpretation at regional and global scales (GLASOD) and more national and quantitative information (ASSOD).

**Soil Organic Matter (SOM):**
Organic carbon content in the upper 100 centimeters of soil
*(metric tons per hectare)*

WISE global data set of soil profiles interpreted for carbon content and soil type (Batjes 1996; Batjes 2000), then applied to FAO's Soil Map of the World mapping units (FAO 1995).

**Soil Nutrient Balance:**
Net nutrient flux *(kg NPK per ha)*

IFDC country- and crop-specific estimates of nutrient balances for Latin America and the Caribbean: mid-1980's and mid-1990's (Henao 1999).

## CONDITIONS AND TRENDS

## INFORMATION STATUS AND NEEDS

♦ Combining the GLASOD soil degradation map with the PAGE agricultural extent suggests that human-induced degradation since the mid 1900s is more severe within agricultural lands. Over 40 percent of the PAGE agricultural extent coincides with GLASOD mapping units that contain moderately degraded areas, and 9 percent of the extent coincides with mapping units that contain strongly or extremely degraded areas. Strong or extreme degradation implies that soils are very costly or infeasible to rehabilitate to their original (mid-1900s) state. This degradation is estimated to have reduced crop productivity by around 13 percent. No global estimates of improving soil quality are known to exist.

♦ Over three quarters of the PAGE agricultural extent contains soils that are predominantly constrained (>70 percent of area has some soil fertility constraints). Just over half the agricultural extent is in flatter lands (up to 8 percent slope). Only 6 percent of land within the PAGE agricultural extent is both flat and relatively free of soil constraints. Most of this land lies in temperate regions.

♦ Depletion of SOM is widespread, reducing fertility, moisture retention, and soil workability, and increasing $CO_2$ emissions. Good land use practices can rebuild SOM levels.

♦ Salinization (an accumulation in the soil of dissolved salts) includes both natural and human-induced (secondary) salinization and occurs on agricultural and nonagricultural, and irrigated and nonirrigated land. Although salinization data is poor, rough estimates indicate about 20 percent of irrigated land suffers from salinization. Around 1.5 million hectares of irrigated land per year are lost to salinization and about $11 billion per year in reduced productivity, or just under 1 percent of both the global irrigated area and annual value of production. Salinization also affects water quality.

♦ Regional analysis of soil nutrient dynamics in Latin America and the Caribbean suggests that for most crops and cropping systems the nutrient balance is significantly negative, although depletion rates appear to be declining (because of substantial growth in fertilizer application in some countries). Previous analyses have demonstrated similar negative balances in Africa.

♦ Relative to its economic and environmental importance, the current lack of comprehensive, reliable, and up-to-date global soil quality data is woeful. Developing reliable, cost-effective methods for monitoring soil degradation is imperative to help to mitigate further losses of productive capacity as well as to target rehabilitation efforts.

♦ The long-term monitoring of SOM and soil biota are increasingly viewed as a strategic means of measuring progress toward achieving sustainable agriculture and of keeping abreast of degradation trends.

♦ Efforts are underway to apply remote sensing methods to soil organic matter monitoring but an adequate ground-truthing capacity still needs to be developed.

♦ National soil research and survey agencies need to strike a more appropriate balance between traditional soil survey activities and fostering scientific and farmer capacity to monitor soil condition on an ongoing basis.

♦ Much more work is needed to develop indicators (and underlying scientific evidence) that relate land use and management, soil quality indicators, production, and economic and environmental outcomes. This is particularly so for better articulating the role of soil quality in providing environmental goods and services from agroecosystems.

♦ Soil nutrient balance data is available at a national level for Latin America and the Caribbean and Africa (Henao 1999; Henao and Baanante 1999) and subnationally for Sub-Saharan Africa (Smaling et al. 1997). Further efforts to make consistent nutrient balance assessments would provide greater insights into the spatial and temporal patterns of agroecosystem productivity.

# Water Services

| **PAGE MEASURES AND INDICATORS** | **DATA SOURCES AND COMMENTS** |
|---|---|
| **Water Supply for Rainfed Agriculture:** Rainfall *(mm per year)* Length of growing period *(LGP in days)* LGP variability | Global estimates of rainfall based on spatial extrapolation of monthly data from rainfall and climate stations over 30 years (UEA 1998). Global length of growing period (LGP) estimates based on University of East Anglia (UEA) data using a water balance model (FAO/IIASA 1999). |
| **Water Use for Irrigation:** Area equipped for irrigation *(percent)* | University of Kassel 1999 global map at half degree resolution (Döll and Siebert 1999);Global Land Cover Characterization Data set (GLCCD 1998; USGS EDC 1999a). |
| Irrigation depth *(mm per year)* | Estimates of national water extraction for agriculture (WRI 1998) and area irrigated (FAOSTAT 1999; Seckler et al. 1998). |
| Efficiency *(ratio of crop water use to amount extracted)* | Country and crop-specific factors (Seckler et al. 1998) applied to the WRI and FAO data, used to determine irrigation depth. |
| **Effects of Agriculture on Water Supply:** Proportion of major watersheds occupied by agriculture | Global digital map of watersheds (UNH – GRDC 1999). |
| **Potential Water Quality Effects:** Soil salinization Fertilizer application rates | Global salinization estimates (Postel 1999; Ghassemi et al. 1995). National fertilizer consumption (FAOSTAT 1999b). |

| **CONDITIONS AND TRENDS** | **INFORMATION STATUS AND NEEDS** |
|---|---|

**CONDITIONS AND TRENDS**

- A fifth of the agricultural extent is in arid or dry semiarid, a quarter in moist semiarid, over a third in subhumid, and a fifth in humid regions. Much of the agricultural extent suffers from high rainfall variability. Global climate change is likely to affect rainfall distribution significantly.
- Irrigation accounts for 70 percent of the water withdrawn from freshwater systems for human use. Of that only 30-60 percent is returned for downstream use, making irrigation the largest net user of freshwater globally.
- The 17 percent of global cropland that is irrigated produces an estimated 30-40 percent of the world's crops. The share of cropland that is irrigated has grown quickly, increasing 72 percent from 1966–1996.
- Competition with other water uses, especially drinking water and industrial use will be most intense in developing countries, where populations and industries are growing fastest. Agriculture may increasingly depend upon water recycled from domestic and industrial uses.
- There is an urgent need to increase irrigation water use efficiency. If efficiency can be improved, less water would need to be extracted from rivers and aquifers per ton of food, feed, or fiber produced. If the excess water is not used to expand food production, improvements in efficiency could help mitigate negative environmental effects of water extraction.
- Over 50 percent of total river basin area is under agricultural cover in the major watersheds of Europe and South Asia; over 30 percent of total basin area is under agricultural cover in large parts of the United States, South America, North Africa, Southeast Asia, and Australia.
- If not carefully managed, agricultural intensification in high external input agroecosystems can result in leaching of mineral fertilizers (especially nitrogen), pesticides, and animal-manure residues into water courses.
- Inadequately managed intensification particularly on more sloping lands with lower quality soils tends to increase soil erosion as well as the effects of sediment on aquatic systems, hydraulic structures, and water usage.

**INFORMATION STATUS AND NEEDS**

- Hydrological monitoring, especially for groundwater levels, river flow, and water quality, is inadequate and, in many cases, has declined in recent years. More cost-effective methods of water resource assessment and hydrological monitoring are needed.
- Even with reliable water quality data it is often difficult to relate water quality changes directly to agricultural activities (e.g., the effects of pesticides from private and public gardens, and the contribution of nutrients from sewage and industrial processing).
- The University of Kassel irrigation map is the best currently available but accuracy is variable because of inconsistent source data (Döll and Siebert 1999). The satellite interpreted Land Cover Characterization Data set (GLCCD 1998; USGS EDC 1999a) is unreliable in detecting irrigated areas, particularly in South America, Africa, and Oceania. Better data is needed on irrigation location, type, and water use (e.g., FAO's AQUASTAT (2000a)).
- Satellite-derived data on rain-use efficiency is now available (University of Maryland 1999) and its suitability as a regional scale indicator of water supply for rainfed agroecosystems merits further investigation. Related indicators of farmer use and management of rainwater need to be developed and tested.

# Biodiversity

| PAGE MEASURES AND INDICATORS | DATA SOURCES AND COMMENTS |
|---|---|
| **Conservation of Natural Habitat:** Proportion of habitat area occupied by agriculture | WWF-US Global 200 Ecoregions (Olson et al. 1999). Major habitats based on broad environmental characteristics and expert opinion. |
| **Pressure on Protected Areas:** Proportion of protected areas occupied by agriculture | Global protected areas database (WCMC 1999). |
| **Habitat Quality in Agricultural Areas:** Proportion of tree cover within the PAGE agricultural extent | Global percentage tree cover data set based on 1 km resolution AVHRR satellite data (Defries et al. 2000). |
| **Agrobiodiversity:** Species Diversity: Crop, livestock, tree crop species | Crop and livestock species (FAO); tree crop species (Kindt et al. 1997). |
| Share of Modern Varieties | Global estimates of the adoption of modern crop varieties (Byerlee 1996; Smale 2000). |
| Germplasm Conservation: Land races and wild species in *ex-situ* and *in-situ* collections. | Data from major germplasm holding institutions: CGIAR China, the United States, and Russia and global estimates of *ex-situ* holdings (Evenson et al. 1998). |
| Area Planted to Transgenic Crops | Percentage estimates for select countries compiled from agricultural survey and census (James 1999). |

## CONDITIONS AND TRENDS

- Agricultural land, which supports far less biodiversity than natural forest, has expanded primarily at the expense of forests.
- About 30 percent of the potential area of temperate, subtropical, and tropical forests has been converted to agriculture.
- Many of the areas established to protect biodiversity fall in or around agricultural lands, increasing the difficulties of effective protection.
- Biodiversity loss is often considerable within high-input agroecosystems, but low-input and extensive systems can also bring about significant biodiversity loss through increased conversion of natural habitats.
- Although tree cover is fairly low in agricultural lands of many parts of the world, a majority of rainfed agricultural land in Latin America, Sub-Saharan Africa, and South and Southeast Asia has significant and increasing tree cover, which enhances habitat for wild biodiversity.
- A number of agricultural systems and management strategies, such as fallowing, agroforestry, shaded coffee, and integrated pest management, can encourage diversity as well as productivity.
- Of the 7,000 crop species used in agriculture, only 120 are important at a national level. An estimated 90 percent of the world's calorie intake comes from just 30 crops. The number of livestock breeds has declined greatly in the past 100 years. The number of domesticated tree crops has increased.
- In the early 1990s, crop area sown to modern varieties of rice and wheat in developing countries had reached around 75 percent, and for maize 60 percent.
- The global area planted with transgenic crops, some 82 percent of which was in OECD countries, increased from only 1.7 million hectares in 1996 to 39.9 million in 1999. The seven principal transgenic crops grown in 1998 were soybean, maize, cotton, canola (rapeseed), potato, squash, and papaya.

## INFORMATION STATUS AND NEEDS

- The currently published national data on land use show *net* land use changes and, therefore, understate the true scale of agricultural conversion impacts on biodiversity. Higher resolution data on land conversion (both spatially and temporally) is needed.
- The size of many protected areas is known but often their precise geographic boundaries are unknown. Increased precision in the delineation of protected area boundaries (as well as in the spatial delineation of agricultural extent) are needed to enhance the quality of the protected area pressure indicator.
- Improved road network and land use/cover data would help determine the level of habitat fragmentation in agricultural landscapes.
- Improved data on production systems diversity could be used as a proxy for the potential area and quality of wildlife habitat within agricultural areas.
- Data on the rate and extent of adoption of modern varieties is relatively sparse, regionally inconsistent, and limited to the major food crops.
- International initiatives are well in hand to georeference germplasm accessions (e.g., WIEWS and ICIS databases). This should considerably enhance agrobiodiversity assays.
- There is insufficient data on the abundance of wild flora and fauna in and around agricultural production areas, and on the impacts of specific crop combinations and management changes on wildlife populations.

# Carbon Services

## PAGE MEASURES AND INDICATORS

**Vegetation Carbon in the PAGE Agricultural Extent:**
Above- and below-ground live vegetation carbon
*(metric tons per hectare)*

**Soil Carbon in the PAGE Agricultural Extent:**
Soil carbon in the first 100cm of the soil profile,
excluding litter *(metric tons per hectare)*

**Carbon-based Greenhouse Gas (GHG) Emissions from Agriculture:**
Carbon dioxide ($CO_2$) emissions
Methane ($CH_4$) emissions

## DATA SOURCES AND COMMENTS

Estimates of carbon density in above- and below-ground live vegetation by ecosystem type (Olson et al. 1983; USGS EDC 1999b) applied to GLCCD (1998) land cover vegetation class extents.

WISE global data set of soil profiles interpreted for carbon content and soil type (Batjes 1996; Batjes 2000), then applied to FAO's Soil Map of the World mapping units (FAO 1995); Latin America and the Caribbean (LAC) example uses the Latin America and the Caribbean Soil and Terrain (SOTER) database (ISRIC 1999; FAO 1998).

$CO_2$ land use change emissions (Houghton and Hackler 1995); $CH_4$ emissions (Stern and Kaufman 1998); Land conversion, livestock, and paddy rice trends (FAOSTAT 1999).

## CONDITIONS AND TRENDS

- Agroecosystems' share of carbon storage is estimated to be 18-24 percent of global total.
- In agricultural areas, the carbon stored in soils is generally more than double that stored in the vegetation that these soils support (102 metric tons/hectare versus 17-47 metric tons/hectare).
- Regional studies for Latin America and the Caribbean show that almost half of the soil carbon is stored in the top 30 cm of the soil profile—the depth most accessible to the rooting systems of annual crops and pasture.
- The primary sources of agriculture-based carbon emissions are biomass burning and methane emissions from livestock and paddy rice production.
- $CO_2$ emissions related to land use change have increased dramatically since the mid 1800s. Most land use change $CO_2$ emissions are now taking place in developing countries.
- Livestock is the largest agriculture-related source of GHG emissions; the growth in livestock populations is also taking place primarily in developing countries.
- Cropland and pasture management strategies that result in improved soil organic matter content also increase carbon sequestration capacity and, thus, help reduce agriculture-induced GHG emissions.

## INFORMATION STATUS AND NEEDS

- Data on carbon storage could be improved by better characterization of agricultural land cover and vegetation carbon content, and by a denser network of soil profile data from agricultural soils.
- Efforts are underway to apply remote sensing methods to soil organic matter monitoring but adequate ground-truthing capacity would still need to be developed.
- It is possible to use satellites to monitor the incidence and severity of fires and hence estimate $CO_2$ and $CH_4$ emissions from biomass burning. However, attributing specific fire incidents to agricultural practices is difficult.
- Monitoring trends in paddy rice areas, ruminant livestock numbers and feed quality, and total livestock provides reasonable proxies for trends in methane emissions from agricultural sources.
- FCCC related research activities into the potential for emission reduction credits from land use change is already generating, and will likely continue to generate, much relevant data and a greater understanding of soil carbon dynamics in agroecosystems.

## Conclusions

As agriculture has become an increasingly dominant influence on global ecosystems, pressures have mounted for agroecosystems to contribute a greater share of society's environmental service needs. However, there are often significant trade-offs between the provision of agricultural and environmental outputs from agroecosystems, holding other conditions constant. Thus, the development of new policies, technologies, and institutional arrangements will be essential if we are to expand the "production possibility frontier" and obtain more agricultural *and* environmental outputs from the world's agroecosystems.

In developing such agroecosystems, society will need to draw on all means at its disposal, including modern biological, agricultural, environmental, and information sciences, as well as the local knowledge of farmers and others. Three broad, interlinked strategies are identified as worthy of further attention in helping to achieve this goal: increasing agricultural productivity (defined using input measures that include natural resource consumption as well as labor, capital, and other purchased inputs); reducing the negative environmental impacts of agriculture; and rehabilitating environmental goods and services within and beyond agroecosystems that would be beneficial for agricultural goods and services.

Current systems of economic valuation fail to reflect not only the long-term value of environmental services from agroecosystems, but often even their current monetary value to users or providers (e.g., increased costs of water purification resulting from agricultural pollution or subsidized provision of irrigation water). New institutional mechanisms are needed to develop effective markets in environmental goods and services. This includes mechanisms to internalize the costs of environmental damage and the benefits of environmental protection into agricultural production and marketing decisions. Such efforts are likely to be most successful where there is a clear, politically expressed perception of environmental scarcity or threat. This will likely happen in areas of population or production pressures, rural and urban poverty, or threatened biodiversity.

Many innovative technologies, systems, institutions, and policies can increase the provision of both agricultural and environmental goods and services from agroecosystems, and improve the well-being of producers and consumers. Only a few, however—such as minimum tillage, organic production of high-value vegetables, integrated pest management, and some agroforestry practices—have been adopted on a regional or global scale. Others are emerging, for example, organic production of traditional export commodities such as coffee and bananas in Costa Rica and Colombia. Greater effort is needed to generate innovations in more environments and farming sys-

tems, to scale up successful strategies, and to rapidly disseminate information on successes and failures.

A dynamic agricultural economy, supportive policies for agricultural development and investment, and strong institutions for information dissemination, research, and marketing are essential, although not sufficient to promote more environmentally friendly production systems. Farmer investment in good land-husbandry practices tends to increase in the following situations: where agricultural markets perform more effectively, reducing the costs of inputs and increasing effective prices received by farmers; where profitable farming opportunities raise the value of agricultural land and water; where technological change makes higher, sustainable yields possible; and where land tenure is secure.

Farming systems, agricultural technology, and the mix of production inputs vary markedly across regions. The availability of land, labor, technology, and capital resources (and, hence, their relative prices) directly affect this variation. Institutions, practices, and technologies required to support more multifunctional agroecosystems will also take different forms in different regions. This presents difficulties for agricultural research and development policymakers. For example, should they spend more on developing technologies that are better adapted to specific local conditions and are likely to impact more limited areas or, on importing and adapting a more broadly applicable solution that may have less impact in any given locale. One of the ways to accelerate the search for appropriate solutions would be to promote institutional mechanisms that foster local screening of new technologies and practices, and whose findings could be fed back into formal technology evaluation and dissemination systems.

In areas with capital intensive production systems and stronger regulatory capacity, such as Europe and North America, it is more feasible to introduce policies, such as taxes and transfers, that internalize major environmental costs. Similarly, improved efficiency in water use could be fostered by promoting water markets. The political challenge lies in convincing producers and consumers to accept such price-increasing policies. There is mounting evidence that consumers are willing to pay for environmental improvement and (real or perceived) improved food safety, especially in more affluent regions of the world. In developing countries where there are more poor people who cannot afford food price increases, and where regulatory frameworks are likely less effective, solutions may lie in more public-investment based options, including technology, infrastructure, and institutional innovation.

Capital intensive farming systems are generally able to maintain and even improve their productive capacity for food and fiber outputs through the use of purchased inputs and capital investments that can ameliorate or compensate for changes in natural resource conditions. Still, such production activities often give rise to negative effects off-site—such as increased water pollution and greenhouse gas emission. Under less capitalized systems, resource degradation has more direct, on-site human welfare consequences. In particular, poor farmers, who rely more on the inherent quality of the resource base, can least compensate for land degradation, loss of wild sources of food, and natural sources of fuel, tools, and building supplies.

The economic process of globalization is fundamentally reshaping the structure of agricultural production and consumption and the scope of environmental policy. Efforts to improve the quality of agroecosystems and, thus, make them more productive, for both agricultural and environmental outputs, must seek out the opportunities that globalization provides. Institutional innovations, such as food-source certification ("organic," "sustainably grown") now make it more feasible for consumers to communicate their demand for agricultural product quality and for the environmental attributes of production systems. And global commitments related to greenhouse gas (GHG) emission control are creating international markets in carbon sequestration and emission reductions that provide incentives for developed country investment in developing countries. More proactive efforts might be required to meet the needs of large agriculture-dependent populations likely to be bypassed by the benefits (if not the risks) of globalization, because of a lack of infrastructure, investment resources, technology, and institutions.

Where proper local support exists, trade offers promising opportunities for farmers to use land and grow products in ways that are resource enhancing. For example, the global diversification of human diets and processing techniques for new food and feed products are expanding markets for tree crops that can be grown in an environmentally sustainable fashion in many tropical and subtropical agricultural regions now considered "marginal." This offers opportunities for employment, economic development, poverty reduction, and improved food security in many poor parts of the world. More proactive efforts are needed to support such diversification, through new technology development for production, processing, storage and use, and promotion of well-functioning market institutions for these new products.

The revolution in communications and information technology should also be harnessed to promote sustainable agriculture. Such applications include: (a) accelerating the flow of information regarding successful technological or institutional innovations, e.g., international science, nongovernmental organization (NGO), and local university use of Internet technologies; (b) improving the resolution, reliability, and availability of satellite data, the growing use of geographical information systems (GIS), and more cost-effective means of handling such geographically referenced data; and (c) integrating advanced

technologies in production systems in ways that both enhance productivity and reduce environmental impacts, such as precision agriculture and drip irrigation.

It is remarkable how much controversy still prevails about the nature, extent, and significance of such key issues as soil degradation, biodiversity loss, and pesticide risks because data are scarce, partial, or too closely linked to advocacy, rather than independent, scientific enquiry. Persistent data gaps limit our ability to monitor the scale and location of environmental problems and successes in the context of agriculture. And significant knowledge gaps, such as in soil biology, limit our ability to design agroecosystem management strategies that enhance positive production system synergies, both biotic and abiotic. More and better-targeted research and development can overcome these weaknesses but would require greater political commitment. Tasks that merit continued public investment include improved satellite monitoring of land cover, soil and water degradation and carbon storage, and the collection of data on land use and resource management practices. There is also a strong case for looking beyond individual ecosystems at cross-ecosystem synergies; for example, the production of biofuels or innovative marine products that have food, feed, or fiber value could reduce food production pressures on agroecosystems, allowing them to contribute more to environmental goods and services. These and many other options are worth further examination by stakeholders in the Millennium Ecosystem Assessment (MEA).

## Recommendations for the Millennium Ecosystem Assessment

1. The science and practice of environmental measurement and valuation in the context of agricultural ecosystems are in their infancy. Development of better methods for spatial, intertemporal, and integrated systems analysis is essential for improved ecosystem assessment and for promoting cost-effective monitoring of the impacts of technological, institutional, and policy change.

2. Fostering the development of agroecosystems that exhibit high levels of agricultural productivity as well as contribute more (or consume less) environmental goods and services will require appropriate policy support. Promising approaches include transfer payments to farmers for environmental services, taxation of agricultural wastes, and transformation of waste products to recycled commodities. Further work is needed to conceptualize alternative policies and document the performance of pilot implementation.

3. The MEA should support international initiatives that seek to advance agricultural and environmental monitoring efforts on a global basis and in spatially referenced formats. The goal should be to help harmonize remote sensing and cross-country survey programs and products, linking them to more detailed local and national monitoring initiatives.

Such information networks should support the capacity to keep abreast of changing natural resource and productivity conditions of the world's major agricultural lands.

4. The collection of remotely sensed and related spatial data is insufficient to interpret changing agroecosystem conditions. Agroecosystems are highly managed, and it is the specific detail of how they are managed that determines their long-term capacity to produce agricultural goods and environmental services. Initiatives are needed to support the standardization and regular compilation of land use and land management data. The most feasible long-term options to collect this type of data probably involve networks operated by and for local communities.

5. The databases, indicators, and collaborator networks developed through the PAGE studies provide a significant resource and should be fully integrated into the MEA activities. One application, for example, might be to link more precise local agroecological and production system characterization into the global-scale schema developed by the PAGE. The PAGE data sets might also provide a sampling framework for stratifying agroecosystem types that are regionally or globally representative and that may serve as foci for organizing MEA activities.

6. This first attempt at evaluating the state of the world's ecosystems was structured according to major biomes: agroecosystems, coastal ecosystems, forest ecosystems, grassland ecosystems and freshwater systems. Many important and often more controversial ecosystem changes occur in the transition areas between ecosystems, such as agricultural productivity in forest margins, and water allocation between agriculture and natural wetlands. We recommend that MEA activities be structured around regional activities that provide incentives to undertake more integrated ecosystem assessments and that seek to better understand, for example, how to meet local goods and service needs by the integrated, or at least harmonized, management of different ecosystems.

# PROLOGUE

The world's agroecosystems provide the overwhelming majority of food and feed on which humanity depends for its continued well-being. The origins of agriculture have been traced to at least 9,000 and perhaps 11,000 years ago in a limited number of regions in which societies domesticated plant and animal species (Smith 1998; Harris and Hillman 1989). As species became increasingly domesticated, such desired traits as increased seed concentration and reduced animal size were enhanced. Various agricultural practices such as seed beds, improved animal nutrition, and water management were also devised. These developments sought to obtain more usable output, such as grain, fruit, meat, and milk, from the selected species as well as to limit losses through pests, diseases, and weed competition (Evans 1998). The increased availability of food, feed, and fiber in turn provided the impetus for societies to prosper and support a larger nonfarming population.

And so it has remained that societies have flourished to a great measure by improving their capacity to expand agricultural production. Globally, agroecosystems have been remarkably successful when judged by their capacity to keep pace with food, feed, and fiber demands. In 1997, agriculture provided 94 percent of the protein and 99 percent of the calories consumed by humans (FAOSTAT 1999).

However, agriculture faces an enormous challenge to meet future food needs. Between 1995 and 2020, global population is expected to increase by 35 percent, reaching 7.7 billion people. An analysis by the International Food Policy Research Institute (IFPRI) suggests that global demand for cereals will increase by 40 percent, with 80 percent of the increase coming from the developing countries. Meat demand is projected to grow by 63 percent and demand for roots and tubers by 40 percent, with 90 percent of this growth coming from the developing world. Demand for fruits, vegetables, and seasonings as well as nonfood farm products will also rise (Pinstrup-Andersen et al. 1999). Although there is still physical capacity to increase production significantly in some of the land-abundant developed countries, limitations of foreign exchange and nonagricultural income sources, as well as high transport and other transaction costs in most developing countries mean that the vast majority of new food supplies will probably need to come from domestic production (McCalla 1999). All regions will face difficulties in meeting the growing demand for agricultural products while also preserving the productive capacity of their agroecosystems.

## Agricultural Intensification

Over most of history—including much of the past century—agricultural output has been increased mainly through bringing more land into production—extending the agricultural frontier through conversion of forests and natural grasslands. But the comparatively limited amount of land well suited for crop production (especially for annual grain crops), the increasingly concentrated patterns of human settlement, and the growing competition from other uses for land have greatly reduced the opportunity for further geographic expansion. Densely populated parts of India, China, Java, Egypt, and Western Europe already reached the limits to geographic expansion many years ago. FAO (1993) estimates that in developing countries between 1990 and 2050, land-person ratios will decline from an average of 0.3 hectares per person to just 0.1-0.2 hectares. This figure will be significantly less in Asia and North Africa.

Intensification of production—obtaining more output from a given area of agricultural land—has become a key development strategy in most parts of the world. In some regions, particularly in Asia, this has been achieved primarily through producing multiple crops each year in irrigated agroecosystems using new, short-duration crop varieties. In high-quality rainfed lands, intensification has been achieved by abandoning fallow periods and modern technologies have made continuous cultivation possible. In many developing countries, there is widespread agricultural intensification on lower quality lands, which are often home to a large proportion of the rural population (Scherr 1999a). There has also been notable intensification of agricultural land use around major cities (and to an unexpected extent, within cities), particularly for high-value perishables, such as dairy and vegetables, but also to meet subsistence needs.

By contrast, in recent decades some developed countries have reverted lower-quality farmland and associated wetlands to more extensive grassland, forest, or conservation uses. This reflects the consequences of market forces (primarily lower real commodity prices), technological development that has favored more intensive use of high-quality lands, and policies designed to retire lower-quality production land—in part to address the growing demand for environmental amenities. Between the early 1960s and the mid 1990s the total amount of agricultural land in Western Europe and North America showed a sustained decline of around 39 million hectares for the first time in modern history (FAOSTAT 1999).

## Environmental Concerns

The unprecedented scale of agricultural expansion and intensification has raised two principal environmental concerns. First, there is growing concern over the vulnerability of the productive capacity of many agroecosystems to the stresses imposed on them by intensification. Local evidence points to such problems as soil salinization caused by poorly managed irrigation systems, loss in soil fertility through overcultivation of the fragile soils of the tropical savannas, and lowering of water tables through overpumping of water for irrigation purposes. But the global significance of this degradation is still far from clear.

Second are the broader concerns about the negative environmental impacts of agricultural production that intensification often accentuates. Such externalities have complex and sometimes far-reaching consequences that must be better understood and more adequately dealt with. Examples include the negative impact of increased soil erosion from hillside farming on downstream fisheries and hydraulic infrastructure, and the damage to both aquatic ecosystems and human health arising from fertilizer and pesticide residues in water sources or on crops. There are also concerns about the loss of habitat and biodiversity from converting land to agricultural uses, as well as narrowing of the genetic base and the genetic diversity of domesticated plant and animal species currently in use. Globally, scientists recognize that agriculture influences climate by altering global carbon, nitrogen, and hydrological cycles.

Different regions, countries, and watersheds around the world with similar rates of agricultural expansion and intensification have experienced different amounts and types of environmental degradation. Some of this variation is due to factors over which farmers have limited influence, such as population growth and density, markets, and the sensitivity and resilience of the natural resource base. But a sizable share of the variation also results from the type and availability of agricultural technologies, natural resource management practices, and the local influence of policies and institutions. Consequently, there is considerable scope for learning from the best (and worst) practices in different places. For example, clear-cutting large expanses of frontier forestland using heavy machinery causes far more biodiversity loss and soil damage than selective, more ecologically sensitive clearing of forest mosaics. Maintaining perennial vegetation in and around cropland may reduce soil erosion, provide habitat for wildlife, reduce forest pressure by providing on-farm wood sources, and improve water percolation and quality. By comparing and contrasting such experiences, one can better understand how some countries have been able to produce similar amounts of agricultural goods and services while giving up less environmental goods and services.

Thus, if point A in Figure 1 approximates the initial endowment of goods and services of two environmentally similar countries, it would be valuable to know how one had been able to follow trajectory A to X, rather than A to Y, in attaining similar levels of agricultural outputs. (Expressed differently, how was one country able to generate more agricultural outputs by giving up the same amount of environmental goods and services following the trajectory A to Z instead of A to X).

## The Policy Challenge

As the world becomes more crowded and as pressures on biological systems and global geochemical cycles mount, it is no longer sufficient to ask whether we can feed the planet. From an agroecosystem perspective, at least, there are more difficult questions. Can the world's agroecosystems feed today's planet and remain sufficiently resilient to feed tomorrow's more hungry planet? Will intensification pressures cause some

*Figure 1*

**Regional Variations in Agriculture-Environment Trade-offs**

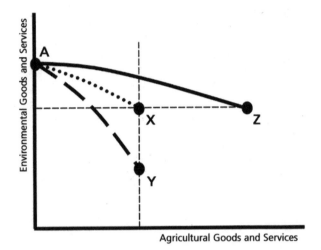

*Figure 2*

**Possible Evolution Paths for Environmental Goods and Services**

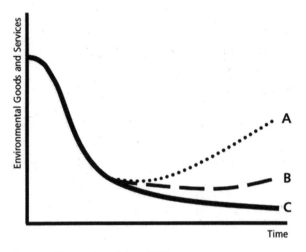

**Source:** Adapted from Scherr 2000.

agroecosystems to irreversibly break down? Presuming we can maintain the productivity of our agroecosystems, are we paying too high a price in terms of the broader environmental effects of agriculture? And, ultimately, how can we improve our policies, institutions, and technologies—at local through to global levels—in ways that maximize the beneficial and minimize the negative consequences of agriculture?

Our spatial analyses make it apparent that agriculture is now a—if not the—dominant influence on the global landscape outside the major urban centers. And policymakers, particularly in densely populated parts of the globe, are finding it increasingly difficult to exempt agroecosystems from playing their role in delivering environmental as well as agricultural goods and services. Thus, at its root, the policy challenge is to foster agroecosystem management practices that will meet growing food, feed, and fiber needs while providing more environmental amenities. This is not just an ethical issue, but one of plain common and economic sense, since the future capacity to deliver agricultural outputs also depends on the continuing ecological viability of agroecosystems.

Policy opportunities to shape future outcomes can arise with respect to both the demand for and supply of goods and services. From a demand perspective, broad development strategies that seek to limit population growth—such as enhanced education (particularly for women), health care, and family planning services—also help limit the growth in aggregate demand for both food and natural resource consumption. Per capita food, feed, and fiber demand will grow as incomes increase, but at a decreasing rate as food becomes a smaller share of household expenditure. Higher incomes are also linked to lower popula-

tion growth rates as well as increased demand for improved environmental amenities.

From a supply perspective, while circumstances in many developing countries often dictate that environmental considerations are afforded less priority by policy makers, continued agricultural intensification need not lead inexorably to environmental degradation. When farmers and farming communities experience environmental degradation, particularly when it affects their own livelihoods, they may respond quite effectively with technical and institutional innovations (*see Figure 2 Trajectory A*). For example, evidence from several parts of the world shows some rural communities have engaged in tree planting to control soil and water movement in farms, regulated cultivation around local water sources, restricted the use of polluting agricultural inputs, and rehabilitated degraded soils. Yet, there is also evidence that Trajectory C is all too common: environmental goods and services—and even the productive capacity for agriculture—have been depleted or as in B, response has been delayed too long for resources to fully recover.

In richer countries, demand and supply priorities for agricultural and environmental outputs are often quite different from those of low-income countries. Low, and even negative, population growth rates and positive, if modest, increases in already high levels of income (on the average and in relative terms) strongly influence the nature of demand. Consumers have pressed for more environmental, particularly recreation-related, goods and services such as more parks, conservation areas, wildlife protection, cleaner waterways and beaches, and so on, and are also expressing increased concern about the *ways* in which food is produced. For example, there is a relatively small, but rapidly growing, demand for organic farm products from consumers willing to pay more for assurances that the use of chemical and biological inputs to production, such as pesticides and animal hormones, is being minimized. This phenomenon could be far reaching as, perhaps for the first time, commercial producers (and the research and development community that support them) are receiving direct market signals to modify production practices in ways perceived to be better from both an environmental and a human health perspective.

Three interrelated strategic objectives merit particular attention in formulating policies to foster the joint expansion of agricultural and environmental goods and services from agroecosystems: increasing agricultural productivity, where productivity is defined using input measures that include natural resource consumption as well as labor and capital; reducing the negative environmental externalities of agriculture; and rehabilitating environmental goods and services within and beyond agroecosystems, such as freshwater, and habitats for plants, pollinators, and predators, that would also benefit agricultural goods and service outputs. Improving agroecosystem management so that all levels of agricultural production can be associated with better environmental performance requires new knowledge and better skills to apply both new and existing knowledge. However, we believe that this can be achieved through improvements in technology, natural resource management systems, landscape planning, and trading opportunities, as well as in the policies and institutional arrangements that help integrate environmental values into agricultural investment and

management decisions. Some real-world examples of progress include the spread of minimum tillage cultivation, integrated pest management, drip irrigation, legume rotations, and live barriers to soil erosion. Expanding the agriculture-environment outputs frontier *(see Figure 3)* over the next 20-50 years represents a major challenge to society globally. For many families in developing countries where poverty is endemic, where population growth is high, and where food security and livelihoods are linked directly to the forces of nature and to environmental goods and services, that challenge is both urgent and daunting.

*Figure 3*

**Enhancing Agroecosystem Goods and Services**

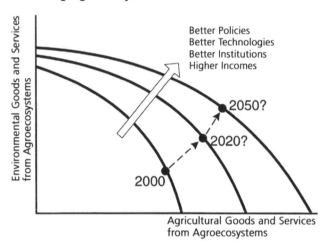

# AGRICULTURAL EXTENT AND AGRICULTURAL LAND USE CHANGES

The first step in assessing the likely global condition of agroecosystems was to determine the location and extent of agriculture. The PAGE study has adopted a common global land cover database for determining the extent of all ecosystems. This chapter describes how we interpret the database in the case of agriculture. The resulting estimate of agricultural extent is compared with higher resolution spatial information as well as with agricultural land use statistics. We also examine the spatial pattern and evolution of major agricultural land use components with the help of statistical data sources. Deficiencies in land use and production system data are then discussed with regard to the challenge of making any meaningful agroecosystem groupings at a global scale. Finally, a highly aggregated agroecosystem characterization schema is defined, based largely on agroecological factors. Wherever possible we use this characterization, or an aggregation thereof, in the remainder of the report to stratify indicators of agricultural and environmental goods and services within the global extent of agriculture.

## Location and Extent of Global Agriculture

The primary data source used in assessing the global extent of agroecosystems was the one-kilometer (1km) resolution, global land cover characteristics database (v1.2) which was initiated by the International Geosphere Biosphere Programme (IGBP) and produced by the Earth Resources Observation System (EROS) Data Center (EDC) of the U.S. Geological Survey and the University of Nebraska-Lincoln (GLCCD 1998; USGS EDC 1999a; Loveland et al. 2000). This data set identifies approximately 200 seasonal land cover regions (SLCRs) per continent (e.g., 167 for South America and 205 for North America) based on interpretation of advanced very high resolution radiometer (AVHRR) satellite imagery consolidated into monthly global composites for the period April 1992 to March 1993. Scientists at USGS EDC and the University of Nebraska-Lincoln established the SLCRs by identifying areas that demonstrated similar land cover associations, physiographic characteristics, dis-

*Table 1*

**PAGE Agricultural Extent Class Descriptions**

| Agricultural Land Cover Class | SLCR[a] Class Types (example of SLCR class) |
|---|---|
| Greater than or equal to 60 percent agriculture | Dominant class is agriculture (Cropland; Cropland *with* ...; Cropland/Pasture) |
| 40–60 percent agriculture | Cropland/Natural Vegetation Mosaic (Cropland/Grassland; Grassland/Cropland; etc...) |
| 30–40 percent agriculture | Dominant class is not agriculture but agriculture is present (... *with* cropland) |
| Other vegetated land cover[b] (0–30 percent agriculture with forest) | Dominant class is forest Agriculture may be present but has not been noted (Forest; Forest with Grassland; etc...) |
| Other vegetated land cover[b] (0–30 percent agriculture with grassland) | Dominant class is grassland Agriculture may be present but has not been noted (Grassland; Grassland with Forest; etc...) |
| Sparsely vegetated | Sparsely vegetated areas (Desert; semidesert; tundra; snow and ice) |

**Source:** IFPRI in consultation with USGS EDC (Brown and Loveland 1998).
**Notes:** (a) SLCR – Seasonal Land Cover Regions of the USGS-EDC produced Global Land Cover Characterization Database v1.2 (GLCCD 1998; USGS EDC 1999a). (b) Other vegetative land cover might contain as much as 30 percent agricultural land, but the actual amount cannot be determined.

tinctive patterns of biomass production, such as the onset, peak, and duration of greenness, and so on (GLCCD 1998; Loveland et al. 2000). The EDC interpretation, thus, attempts to capture both spatial and seasonal variations in vegetation cover.

The classification system used in the original EDC data set gave some scope for refining the data interpretation for agricultural purposes, e.g., to improve upon the IGBP interpretation of land cover that is generally presented at a global level (IGBP 1998). Previous land cover interpretations made from the data set had not explicitly recognized all occurrences of agriculture occupying a less than dominant share (60 percent) of a SLCR class. In consultation with EDC, the potential agricultural content of all 961 SLCRs defined globally was reassessed (Brown and Loveland 1998). For example, an area interpreted as containing more than 60 percent forest and classified as, "deciduous broad-leaf forest" cover using the IGBP classification scheme might contain an agricultural subcomponent, e.g., its detailed classification might describe it as *"deciduous broad-leaf forest with cropland."* The reassessment aimed to identify all such occurrences of agriculture, even when they occurred as minor cover components, although this was limited by the seasonal land cover naming system, which did not identify an agricultural component if it occupied less than 30 percent of the land cover class area. Nevertheless, the reassessment did define three primary agricultural cover categories based on ranges of agricultural area intensity (30-40, 40-60, and greater than 60 percent agriculture), and two indeterminate categories

in which agriculture might feasibly occur but whose area intensity lay below the SLCR threshold of 30 percent *(see Table 1).* The new interpretation of the global location and extent of agriculture is shown in Map 1.

The reinterpretation alone, however, does not address some intrinsic weaknesses in the land cover characterization data set, such as regional variations in the reliability of the satellite data interpretation, reflecting differences in the structure of land cover and the availability of reliable ground-truthing data (Brown and Loveland 1998). Specific agricultural landcover types whose interpretation is considered to be problematic include: extensive dryland arable farming; irrigated areas; permanently cropped areas (especially tree crops in forest margins); and extensive pasture land.

Taken at its face value, Table 2, derived from the extent analysis that produced Map 1, suggests that Latin America and the Caribbean, Sub-Saharan Africa, and the Former Soviet Union have the largest amount of agricultural land, each making up around 16 to 17 percent of the PAGE global agricultural extent. Europe, South Asia, and Southeast Asia, however, have the highest proportion of their total land area under agriculture (71 percent, 73 percent, and 47 percent, respectively), while agriculture in Oceania and West Asia/North Africa is concentrated into less than 10 percent of the total land area. Both South Asia and Southeast Asia have around 70 percent of their agricultural extent in the most area intensive agricultural use (60 percent or greater under agriculture) *(see Figure 4a).*[1] Globally,

*Table 2*

## PAGE Agricultural Extent and FAO Agricultural Land by Region

| Region | PAGE Agricultural Extent, 1992-93 | | | FAO Agricultural Land, 1992-93 Average Area | | | | | | | | |
| --- | --- | --- | --- | --- | --- | --- | --- | --- | --- | --- | --- | --- |
| | | | | Total Agriculture[b] | | Crop-land[c] | Annual Crops[d] | | Permanent Crops[e] | | Permanent Pasture | Irrigated Land |
| | Total Land Area[a] | Area | Share of Total Land Area | Area | Share of Total Land Area | Area | Area | Harvested Area | Area | Harvested Area | Area | Area |
| | (000 sq km) | | (percent) | (000 sq km) | (percent) | --------------------------------(000 sq km)-------------------------------- | | | | | | |
| North America | 18,722 | 4,406 | 23.5 | 4,991 | 26.7 | 2,306 | 2,284 | 1,223 | 22 | 17 | 2,685 | 221 |
| Latin America and the Caribbean | 20,178 | 6,254 | 31.0 | 7,552 | 37.4 | 1,531 | 1,279 | 924 | 251 | 154 | 6,004 | 171 |
| Europe | 4,726 | 3,336 | 70.6 | 2,156 | 45.6 | 1,365 | 1,227 | 837 | 138 | 130 | 791 | 167 |
| Former Soviet Union | 17,877 | 5,716 | 32.0 | 5,863 | 32.8 | 2,299 | 2,246 | 1,288 | 53 | 35 | 3,555 | 204 |
| West Asia/ North Africa | 11,890 | 1,135 | 9.5 | 3,682 | 31.0 | 920 | 816 | 571 | 104 | 92 | 2,762 | 239 |
| Sub-Saharan Africa | 22,676 | 5,837 | 25.7 | 9,909 | 43.7 | 1,659 | 1,465 | 1,207 | 193 | 156 | 8,254 | 62 |
| Asia | 24,703 | 8,859 | 35.9 | 10,304 | 41.7 | 4,443 | 3,975 | 4,168 | 468 | 380 | 5,861 | 1,434 |
| East Asia | 16,213 | 3,792 | 23.4 | 6,995 | 43.1 | 1,503 | 1,407 | 1,506 | 95 | 100 | 5,492 | 580 |
| South Asia | 4,129 | 3,028 | 73.3 | 2,231 | 54.0 | 2,035 | 1,946 | 2,023 | 89 | 106 | 196 | 712 |
| Southeast Asia | 4,360 | 2,039 | 46.8 | 1,078 | 24.7 | 905 | 621 | 640 | 284 | 174 | 173 | 142 |
| Oceania | 8,491 | 690 | 8.1 | 4,826 | 56.8 | 518 | 492 | 167 | 25 | 11 | 4,308 | 24 |
| Global Total | 130,484 | 36,233 | 27.8 | 49,281 | 37.8 | 15,039 | 13,785 | 10,386 | 1,254 | 975 | 34,219 | 2,523 |

**Source:** PAGE Agricultural Extent: IFPRI reinterpretation of GLCCD 1998; USGS EDC 1999. FAO Agricultural Land: FAOSTAT 1999. Regional boundaries: ESRI 1996.

**Notes:** (a) Total land area is based on FAOSTAT 1999. (b) FAO total agricultural land includes cropland and permanent pasture. (c) FAO cropland is annual crops plus permanent crops, including irrigated areas. (d) Annual crops include land under temporary crops (double cropped areas are only counted once), temporary meadows for mowing or pasture, market and kitchen gardens, and land temporary fallow (less than five years). (e) Permanent crops include land cultivated with crops that occupy the land for long periods and need not be replanted after each harvest, but excludes land under trees grown for wood or timber (FAOSTAT 2000).

land with greater than 60 percent agriculture occupies 41 percent of the PAGE agricultural extent, while land with 40-60 percent and 30-40 percent agriculture occupies 29 and 30 percent of the PAGE agricultural extent, respectively.

Some of the most intensively managed agricultural lands are located in the most densely settled areas, notably in Western Europe, India, eastern China, and Java. Overall, nearly three quarters of the world's total population live in what is defined as the PAGE global area extent of agriculture (IFPRI calculation using CIESIN 2000). Because the Seasonal Land Cover Regions do not explicitly recognize an urban class, the PAGE global extent of agriculture includes urban areas. The calculation of populations residing within the PAGE extent of agriculture, therefore, includes both rural and urban populations. If the population for urban areas, as defined by the Stable Lights and Radiance Calibrated Lights of the World (1994-95) data base (NOAA-NGCD 1998; Elvidge et al. 1997), is excluded, the percentage of total population living within the extent of agriculture drops to about 50 percent of the world's total population (IFPRI calculation using CIESIN 2000).

## COMPARISON WITH OTHER DATA SOURCES

The most frequently employed spatial reference for the global extent of agriculture is the IGBP Land Cover Classification map (IGBP 1998). Relative to the IGBP classification, the PAGE classification expands the geographic extent of agriculture by including areas where agriculture is not the dominant land cover and discriminates three agricultural area intensity classes. A comparison found that the PAGE global agricultural extent includes all of the IGBP cropland and cropland/natural vegetation mosaic areas and an additional 6 percent of the IGBP forest area and 14 percent of the IGBP grassland and shrubland area.

To illustrate the likely discrepancies between global and regionally-derived spatial information, Map 2 compares the PAGE agricultural extent for Central America with another developed by the Centro Internacional de Agricultura Tropical (CIAT) for the same region using a compilation of more detailed information sources, including national agricultural census data (Winograd and Farrow 1999). The area is predominantly humid tropics with heterogeneous cover, such as mixed forest, permanent crops, pasture, and cropland. Although the maps gener-

*Figure 4*

## a. Composition of PAGE Agricultural Extent by Region, 1992–93

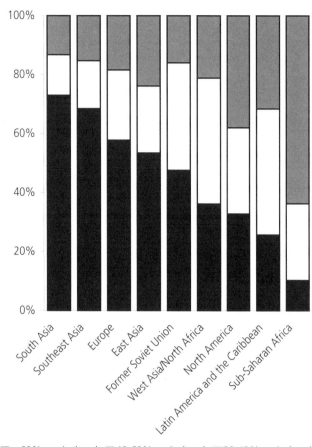

■ > 60% agricultural   □ 40-60% agricultural   ▨ 30-40% agricultural

**Source:** IFPRI reinterpretation of GLCCD 1998; USGS EDC 1999a.

## b. Composition of FAO Agricultural Land by Region, 1992–93

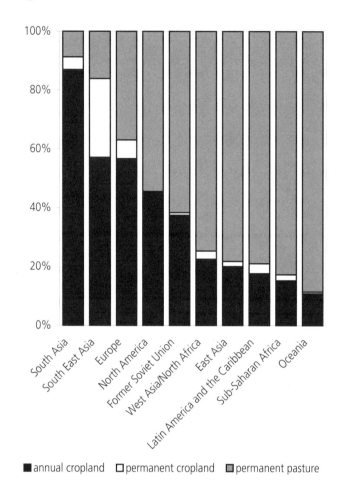

■ annual cropland   □ permanent cropland   ▨ permanent pasture

**Source:** FAOSTAT 1999.

ally point to the same broad tracts of land as being agricultural, there are many detailed and several major differences in the agricultural extent they define. Some of these differences may be attributable to different time frames—the PAGE map draws on satellite imagery from a single year 1992-93, whereas the CIAT map is a composite of finer resolution data from a range of years across the 1980s and 1990s and includes some modeled allocation of agricultural areas guided by road proximity.

At the global and regional scale, we can also compare the satellite estimates of agricultural extent with the agricultural land use statistics for 1992-93 compiled by FAO *(see Table 2 and Figure 4b)*.[2] FAO reports agriculture, cropland plus permanent pasture, to be around 37 percent of global land during this period, over 30 percent greater than the satellite estimate (FAOSTAT 1999). There are at least two sources of discrepancy. The first is the under reporting of agricultural area by satellite remote sensing because of both the failure to detect

agricultural crops and pasture that are similar to natural forests, woodlands, and grassland, and the limitation of not identifying agriculture when it occupies less than 30 percent of a land cover class. The second is in treating the FAO cropland and pasture statistics as measures of the *actual* amounts of land in use in any year when they include temporary fallow areas (FAOSTAT 2000). Fallow areas were most likely not detected as agricultural using the satellite imagery. During the period corresponding to the satellite data collection (1992-93), FAO reports global cropland as 1.5 billion hectares but total harvested crop area as only 1.1 billion hectares. Furthermore, harvested area, itself, is often greater than the *physical* crop area, as is potentially detectable by satellite, because of multiple cropping in many parts of the world, particularly in irrigated areas.

These comparisons suggest the need for a conservative approach to applying and interpreting all forms of aggregate land cover and land use data. Indeed, persistent problems of obtain-

*Table 3*

## Trends in Global Agricultural Land Use

| | 1965–67 | | 1975–77 | | 1985–87 | | 1995–97[a] | | Annual Growth Rates | | |
|---|---|---|---|---|---|---|---|---|---|---|---|
| | Area | Share of Agriculture | Area | Share of Agriculture | Area | Share of Agriculture | Area | Share of Agriculture | 1966–76 | 1976–86 | 1986–96[a] |
| | *(Mha)* | *(percent)* | *(Mha)* | *(percent)* | *(Mha)* | *(percent)* | *(Mha)* | *(percent)* | *(percent per year)* | | |
| Cropland | 1,365 | 30.0 | 1,411 | 30.2 | 1,482 | 30.7 | 1,508 | 30.6 | 0.33 | 0.49 | 0.18 |
| Annual Crops | 1,280 | 28.1 | 1,315 | 28.2 | 1,374 | 28.4 | 1,378 | 28.0 | 0.27 | 0.44 | 0.03 |
| Harvested | 940 | 20.7 | 1,002 | 21.5 | 1,030 | 21.3 | 1,063 | 21.6 | 0.64 | 0.28 | 0.32 |
| Permanent Crops | 85 | 1.9 | 96 | 2.1 | 108 | 2.2 | 130 | 2.7 | 1.24 | 1.21 | 1.88 |
| Harvested | 61 | 1.3 | 71 | 1.5 | 89 | 1.8 | 104 | 2.1 | 1.50 | 2.26 | 1.56 |
| Permanent Pasture | 3,182 | 70.0 | 3,259 | 69.8 | 3,353 | 69.3 | 3,418 | 69.4 | 0.24 | 0.28 | 0.28 |
| Irrigated Land | 153 | 3.4 | 192 | 4.1 | 226 | 4.7 | 264 | 5.4 | 2.28 | 1.65 | 1.57 |
| Total Agriculture | 4,597 | 100.0 | 4,670 | 100.0 | 4,835 | 100.0 | 4,924 | 100.0 | 0.27 | 0.35 | 0.26 |
| Global Total[b] | 13,044 | 34.9 | 13,044 | 35.8 | 13,044 | 37.1 | 13,048 | 37.7 | | | |

**Source:** IFPRI calculation based on FAOSTAT 1999.
**Notes:** (a) FAO does not report agriculture area and permanent pasture after 1994, therefore, the averages under the 1995–97 column for these two items are the average of 1992–94. Accordingly, the growth rates under column 1986–96 for agricultural area and permanent pasture are calculated using 1986–1992. (b) The share value for global total is total agriculture as a percentage of total land.

ing reliable and comparable estimates has lead FAO to discontinue reporting permanent pasture and forest area data in its land use statistics since 1994. Young (1998) provides further examples of the weaknesses of existing land use statistics.

For the purposes of the PAGE study, we interpret the agricultural extent depicted in Map 1 as a broad expression of the area occupied by more spatially intensive forms of agriculture (greater than 30 percent agriculture by area) between May 1992 and April 1993. Therefore, it most reliably represents land occupied by annual crops and, with some regional variability, perennial crops and more highly managed pastures.

## Agricultural Land Use Balances and Trends[3]

The past century has seen unprecedented growth in agricultural expansion, reflecting rapid population growth, generally rising—though still highly variable—standards of living, market integration, urbanization, new technologies, and agricultural investment. Agricultural expansion, and the subsequent intensification of production methods as good arable land became more scarce, brought about impressive increases in food availability and profound effects on natural habitat. Turner et al. (1990) estimated that the global extent of cropland increased from around 265Mha in 1700 to around 1.2Bha in 1950, predominantly at the expense of forest habitats *(see Biodiversity section)*, and now stands at around 1.5 Bha. Conversion to permanent pasture, from both grassland and forests has been even more expansive, reaching around 3.4 Bha in the mid 1990s.

Only limited areas remain totally unaffected by agriculture-induced land use changes, even though our spatial analysis suggests that only around 30 percent of total land area is in agriculture-dominated landscapes or agricultural mosaics. Global land-use patterns and trends over the last several decades are summarized in Table 3.

### CROPLAND AND PASTURE

Aggregate land-use statistics reveal that around 31 percent of the global extent of agriculture is crop based, with the remaining 69 percent in pasture. About 91 percent of cropland is occupied by annual crops, such as wheat, rice, and soybean, with the balance being used for the cultivation of permanent crops, such as tea, sugarcane, coffee, and most fruits. However, these global averages mask large regional differences, as shown in Figure 4b. In some regions pasture-based agroecosystems predominate: 89 percent in Oceania, 83 percent in Sub-Saharan Africa, 79 percent in Latin America and the Caribbean, and 78 percent in East Asia. In other regions, crop-based agroecosystem shares are much larger: 91 percent in South Asia, and 84 percent in Southeast Asia. India has a staggering 94 percent of its total agricultural extent under crops. In regions where arable agriculture predominates, there are also higher proportions of permanent crops, but the 32 percent of permanent cropland in Southeast Asia far outstrips the next most important regions: Western Europe and South Asia.

### IRRIGATION

Globally, around 5 percent of agricultural land (264 million hectares) is irrigated, but that share stands at 35, 15, and 7

*Figure 5*

## Irrigation Intensity, 1995–97

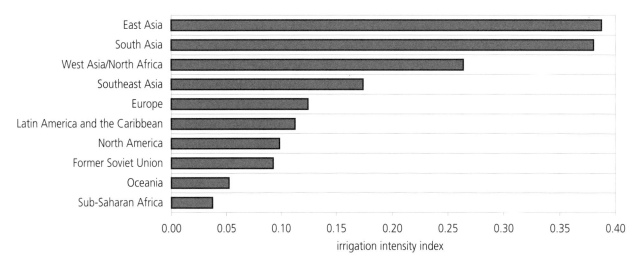

**Source:** Compiled from FAOSTAT 1999.
**Note:** The irrigation index is the irrigated area divided by the total area of cropland.

percent respectively of the agricultural lands in South Asia, Southeast Asia, and East Asia. China and India alone contain 39 percent of the global irrigated area, and Western Europe and the United States contain another 13 percent. Sub-Saharan Africa and Oceania, on the other hand, have less than one percent of their agricultural land irrigated. Considering irrigated area as a share of annual cropland (the primary use for irrigation water in most regions), the perspective changes only slightly. The global ratio of irrigated area to annual crop area is 21 per-

cent but that ratio reaches as high as around 40 percent in South and East Asia and as low as 4 percent in Sub-Saharan Africa and 5 percent in Oceania, as shown in Figure 5.

### CROPPING INTENSITY

A simple measure of the intensity of both irrigated and rainfed agriculture is cropping intensity, defined here as annual harvested area as a proportion of total cropland (land in use plus fallow). For example, swidden agriculture relies on maintaining

*Figure 6*

## Annual Cropping Intensity, 1995–97

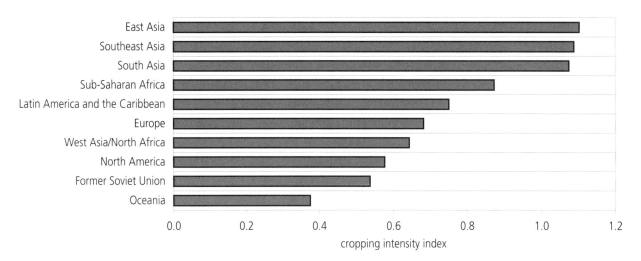

**Source:** Compiled from FAOSTAT 1999.
**Note:** The cropping intensity index is the harvested area of annual crops divided by the total area of annual cropland.

a significant share of production area in fallow every year, thus, having a low cropping intensity (less than 1); whereas some irrigated areas that can produce up to three crops per year from the same physical area, thus, have a cropping intensity of 3. On a global basis, the average cropping intensity for annual crops is about 0.8 but regional variations are again significant *(see Figure 6)*. Areas with high levels of irrigation, such as South Asia, have average intensities of 1.1. Asia as a whole has an annual cropping intensity of 1.0. India, China, and Indonesia play a large part in shaping these totals with their annual cropping intensities of 1.1-1.2. These intensities are in sharp contrast to temperate, developed-country agriculture, such as Western Europe and North America, where temperature-limited growing seasons and proportionately less irrigation give rise to annual cropping intensities of 0.6-0.7. The intensity of 0.9 for Sub-Saharan Africa is surprisingly high and implies a greater intensity of farming than expected of a region with little irrigation and the common use of fallow periods. But this finding might simply reflect data weaknesses, in this case associated with the under reporting of agricultural lands, where fallow lands, land farmed under subsistence crops, and crops that grow within forested areas are often not correctly accounted for in agricultural land use statistics.

## AGRICULTURAL LAND USE TRENDS

Globally, the net rate of growth of agricultural land has been relatively low and constant at around 0.3 percent per annum for each of the last three decades: 1966-76, 1976-86, and 1986-96 *(see Table 3)*. The total area of agriculture has risen from 4.6 billion hectares in 1966 to 4.9 billion hectares in 1996. During the first two decades, the growth in annual cropland exceeded the growth of permanent pasture, but the period 1986-96 witnessed a general decline in cereal area and annual cropland remained fairly static overall, while permanent pasture continued to grow at its former rate. The two most dynamic components of agricultural land use since the early 1960s have been expansion in permanent crops from 85 million hectares (about 2 percent of agriculture) to 130 million hectare in 1996 (almost 3 percent of agriculture) and the progressive, but slowing, growth in irrigated area from 153 million hectares in 1966 to 264 million hectares in 1996.

Regional trends, again, paint a more dynamic picture. A highly significant trend is the decrease in land devoted to agriculture in some developed countries. Both Western Europe and the United States have progressively been taking land out of agriculture over the last 40 years and Oceania, to a lesser extent, for the last 25. These three regions have retired around 41 million hectares from agricultural production since 1966, although most recent statistics indicate some reversal of this trend since the mid-1990s. There has also been a significant, although perhaps more temporary, reduction in total agricultural land in

Eastern Europe, attributable to the liberalizing of production and marketing regimes and the recent poor economic conditions of those countries. South Asia's total agricultural area has remained at around 223 million hectares for over 20 years. Across all regions, West Asia is the only region in which total agricultural land is still growing at greater than 1 percent per year from 1986-96. Figure 7 shows the trends in primary agricultural land-use components for four regions, highlighting the significant differences in agricultural evolution since the early 1960s. A notable common feature has been growth in the irrigated proportion of agricultural land, otherwise, trends in cropland and cropping intensities for annual and permanent crops vary significantly.

## LAND PRODUCTIVITY IMPLICATIONS

Even the modest growth in the global amount of agricultural land (0.3 percent per annum) translates into an additional 12.5 million hectares a year (an area the size of Greece or Nicaragua) in net expansion. The data also imply that agricultural land productivity must, on average, have increased given that growth in both population (1.8 percent per annum) as well as per capita food consumption (0.6 percent per annum) expanded total food consumption at a consistently faster rate than agricultural land expansion over the same period. The data also show that part of the productivity improvement has come from increased production intensity, as revealed here by the higher growth in irrigated land within agricultural land than agricultural land itself, and the continuous increase in aggregate cropping intensity.

## Agricultural Land Use Dynamics

As is evident in comparing global and regional land use trends, aggregate statistics mask important geographical variations. Although the global growth rate of agricultural land between 1986 and 1996 was around 0.3 percent per year, it ranged at the regional level from 1.4 percent per year in West Asia to –0.6 percent per year in Western Europe. Similarly, global annual crop area was more or less constant over the last 10 years, while 2 of the larger countries, agriculturally—Brazil and Indonesia—have witnessed growth rates in annual crops of 2.2 and –1.4 percent per year respectively. Hence, at each geographic level of disaggregation, the dynamics of land use change exhibit greater variation.

From an ecosystem perspective, there is an even greater limitation inherent in such land use statistics. Regardless of the geographical scale, these statistics report only *net changes* in land use. Consider a country in which for any given year 200,000 hectares of forest were cleared for agriculture and 200,000 hectares of degraded agricultural land were reforested. At year end the national aggregate statistics show the same net amount of agricultural land—apparently zero change. But, from an eco-

*Figure 7*

## Land Use Trends in Selected Regions

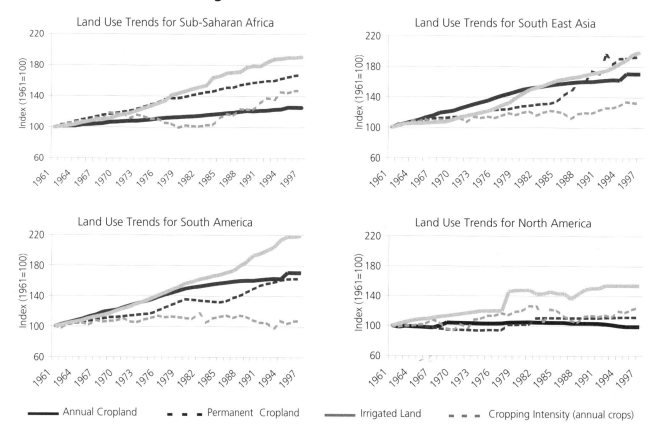

**Source:** IFPRI calculation based on FAOSTAT 1999.

system perspective, important transitions would have occurred on 400,000 hectares of land. This deficiency calls for greater emphasis to be placed on full land use accounting, a problem that the Kyoto Protocol may help to address through incentives created for monitoring the carbon sequestration and emission impacts of land use change *(see Carbon Services section).*

## Agroecosystem Characterization within the PAGE Global Agricultural Extent

We define an agroecosystem as *"a biological and natural resource system managed by humans for the primary purpose of producing food as well as other socially valuable nonfood products and environmental services."* Although an agroecosystem may be functionally equivalent to a farming system or a land utilization type (Conway 1997; FAO 1981), our definition emphasizes the interplay of biotic and abiotic resources in and around farmed areas and the potential to manage agroecosystems for the generation of a broader range of goods and services than food. Agroecosystems are enormously diverse, but the set of economically feasible production options at any given location

is generally conditioned by agroecological factors: radiation, temperature, rainfall, slope, soil fertility, endemic pests and diseases, and so on. In each agroecological domain, socioeconomic factors affect the human selection and management of specific productive activities. These include: the existence of and access to physical infrastructure and markets; population pressure; consumption preferences; land use and natural resource property rights; access to technologies, information, and credit; and cultural norms.

The most appropriate agroecosystem concept for the PAGE study would have been to identify frequently occurring agroecological and socioeconomic patterns that have given rise to globally significant livelihood strategies such as: smallholder, rainfed, paddy-rice systems in the humid tropics; commercialized, rainfed, cereal and oil-crop systems in the subhumid and humid temperate zones; and subsistence food-crops or cattle ranching in the forest margins of the humid tropics and subtropics. Unfortunately, there is little information on actual land use and land management, even broadly defined, above local or watershed level, and certainly no comparable geographically-referenced data at the international level. The land cover characterization database contains some generalized but inconsistently applied details of agricultural land cover types while, on the other hand, FAO's rich national statistics on land use and crop and livestock production (FAOSTAT) give no clue as to geographic distribution within countries. There are a number of regional or global crop distribution maps (for example CIAT 1996; CIP 1999; and USDA 1994) but it was not possible to compile, harmonize and integrate such data for the PAGE study. Furthermore, crop distribution alone says nothing about *how* the crops are produced. It is precisely this information that is most vital when considering the environmental dimensions of agroecosystem condition. Using the same geographical example of Central America as depicted in Map 2, Maps 3a and 3b show the type of information that would be useful on a regional basis to make an improved global scale agroecosystem assessment. Map 3a depicts the major production system mapping units of Map 2 further disaggregated by agroecological characteristics, and Map 3b shows a stratification of dominant agroecosystems by the general level of intensification being practiced (Winograd and Farrow 1999).

Within the boundaries of the data constraints, but with the goal of providing a global, ecosystem-oriented perspective on agriculture, we developed an aggregated agroecosystem characterization schema for the PAGE study. The purpose of the characterization schema is to provide an ecosystem-oriented spatial framework (in addition to regional and geopolitical units) for aggregating agricultural and environmental indicators within the PAGE global extent of agriculture. We constructed the spatial stratification schema, described in this section, using global agroecological data generated by FAO and IIASA (FAO/IIASA 1999; IIASA 1999) and a global irrigated area map generated by the University of Kassel (Döll and Siebert 1999).

## AGROCLIMATIC FACTORS

Over the past 25 years, FAO has developed both the conceptual elements and the global databases to summarize biophysical information, such as climate, slope, and soil, into a small number of composite "agroecological zone" (AEZ) variables that are significant for farm management, and crop and pasture productivity (FAO 1978-81). The University of East Anglia (UEA 1998) generated the underlying climate data used in deriving the most recent FAO/IIASA global AEZ data set (FAO/IIASA 1999). The data includes 30 years (1961-90) of monthly climate data covering the world's landmass at a 0.5 degree grid resolution. The AEZ variables built from this data set and used in the PAGE study (IIASA 1999) are the following:

### Major climate
♦ Tropical, subtropical, or temperate class, and subdivisions indicating, for example, summer or winter rainfall patterns.

### Length of growing period
♦ The average length of growing period (LGP, in days per year) from an analysis of 30 years of monthly data for each grid cell.
♦ The coefficient of variation of the LGP for each grid cell based on the 30-year monthly analysis.
♦ The average temperature during the growing period.

### Thermal zone
♦ Combining the major climate variable with the average temperature during the growing period to create a thermal zone variable.

The spatial correspondence of distinct thermal zones and LGP ranges are termed *agroclimatic zones*.[4] The LGP is the period when both thermal and moisture conditions are suitable for crop growth. Moisture availability constrains the growing season of most rainfed agroecosystems in tropical and subtropical areas, whereas temperature and moisture jointly constrain the growing season in most temperate areas.[5]

The global agroclimatic map *(see Map 4)* delineates domains that have roughly the equivalent potential for rainfed agriculture, from a climatological perspective, within the PAGE global extent of agriculture. Table 4 provides a tabular summary of PAGE global agricultural extent by the major climatic zones.

Within the PAGE global extent of agriculture, some 38 percent is in tropical, 23 percent in subtropical, and 38 percent in temperate environments. Nearly 10 percent of agricultural lands are located in arid regions, made possible only through irrigation; and nearly half of these are in temperate climates. The most prominent agroclimatic zones for agriculture are the subhumid temperate, the subhumid tropics, and the humid tropics (16, 15, and 9 percent, respectively) which, together with the dry semiarid tropics and sub-tropics and humid temperate areas, account for about two thirds of the agricultural extent.

## GENERALIZED SLOPE

In addition to climatic conditions, other  biophysical factors, such as soils and topography, are important determinants of agroecosystem capacity at the global and regional scale. Because of their spatial complexity, soils are omitted from the PAGE global scale stratification but are analyzed separately as determinants of both agroecosystem potential and condition in the section on Soil Resource Condition. We represent topography by a generalized slope variable. Slope has several important influences on the management of agroecosystems. For example, flat lands are more suited to irrigation and mechanized production but can present waterlogging problems, while sloping lands are likely to be better drained but more vulnerable to soil erosion. The generalized slope variable is part of the FAO/IIASA global AEZ database (IIASA 1999) and is derived from the glo-

*Table 4*

## PAGE Agricultural Extent by Agroclimatic Zone and Slope Class

| Agroclimatic Zone[a] | Land Area Within the PAGE Agricultural Extent[b] | Share by Agroclimatic Zone | Share by Slope Class[c] | | | | |
|---|---|---|---|---|---|---|---|
| | | | Flat | | Moderate | Steep | |
| | | | 0–5 | 5–8 | 8–16 | 16–30 | >30 |
| | *(m sq km)* | | *(percent)* | | | | |
| Tropics/Arid and Semiarid | 5.2 | 14.4 | 3.5 | 4.8 | 3.5 | 1.8 | 0.7 |
| Tropics/Subhumid and Humid | 8.5 | 23.5 | 7.9 | 5.4 | 4.4 | 3.7 | 1.7 |
| Subtropics/Arid and Semiarid | 3.4 | 9.4 | 3.6 | 1.8 | 1.4 | 1.4 | 1.2 |
| Subtropics/Subhumid and Humid | 5.0 | 13.8 | 3.0 | 2.2 | 2.4 | 3.0 | 3.0 |
| Temperate/Arid and Semiarid | 7.3 | 20.1 | 5.6 | 6.2 | 4.0 | 2.9 | 1.5 |
| Temperate/Subhumid and Humid | 6.5 | 18.0 | 4.5 | 5.6 | 3.8 | 2.9 | 0.9 |
| Boreal | 0.3 | 0.8 | 0.1 | 0.2 | 0.3 | 0.2 | 0.0 |
| Total | 36.2 | 100.0 | 28.2 | 26.2 | 19.9 | 15.9 | 8.9 |

**Source:** IFPRI calculation based on: (a) FAO/IIASA 1999; (b) GLCCD 1998; USGS EDC 1999a; and (c) FAO/IIASA 1999 based on USGS 1998.

bal United States Geological Survey (USGS) digital elevation data set (USGS 1998).

Table 4 reports the percentage of the PAGE agricultural extent within each slope class and agroclimatic zone. Globally, 54 percent of the agricultural extent can be considered "flat" (under 8 percent slope) and 20 percent is located on moderate slopes (8-15 percent). A quarter of the agricultural extent has steep slopes (16 percent or more), a third of which are very steep (over 30 percent slope). About 3.2 million km[2] of steeply sloping farmland is found in agriculture-dominant areas (greater than or equal to 60 percent agriculture) and 5.7 million km[2] in agriculture mosaics (30-60 percent agriculture). All climatic zones have a similar proportion, about 8 percent, of agricultural extent with steep slopes. The unexpected prevalence of temperate agriculture on steep slopes is due to the amount of agricultural land in central Europe, the Himalayas, and northern China. These figures somewhat overestimate the importance of steeplands, as it is likely that agricultural fields are concentrated on less steep niches within the agricultural extent.

### IRRIGATION

A rare example of a global-scale spatial data set that provides a direct insight into agricultural land management practices is the map of areas equipped for irrigation generated by the University of Kassel (Döll and Siebert 1999). This is the first published digital map resulting from ongoing work to compile, harmonize, and integrate irrigation data in spatial formats. The irrigation area map *(Map 16)* appears in the Water Services section.

For the purpose of globally characterizing agroecosystems, we grouped the Kassel data on percentage of land equipped for irrigation into three ranges. We designated *irrigated* lands as

those within which at least 15 percent of the area was equipped with irrigation infrastructure. This excludes many areas of tube well, hose, and simple ditch irrigation constructed at the farm level. However, it does include irrigated areas that lie outside the PAGE satellite-defined agricultural extent because they had less than 30 percent agricultural cover.[6] Those lands that contained between 5-15 percent of area equipped for irrigation were classed as *mixed irrigated/rainfed.* In these areas, irrigation probably significantly influences the agricultural economy and ecology, but there is still a great dependency on rainfed production. In dry areas (defined as those with under 120 growing days), irrigation is often essential for production, while in wetter areas it may be supplemental. Overall, *irrigated* and *mixed irrigated/rainfed* lands make up 15 percent of the extended global agricultural extent. We defined areas containing less than 5 percent of area equipped for irrigation as *rainfed* lands.

### GLOBAL AGROECOSYSTEM CHARACTERIZATION

We derived the aggregate agroecosystem characterization schema by combining the agroclimatic, slope, and irrigated area data themes with that of the PAGE global extent of agriculture. Although this is far from a true typology of agroecosystems, it does identify domains within which agricultural systems and practices are likely to have common features and to face similar environmental constraints and opportunities. This characterization does not address variation in production intensity, nor the combination of annual and perennial crops and livestock components. To keep the number of global agroecosystem groupings in a manageable range (no more than 20), it was necessary to reduce the number of classes of some variables. Furthermore, while it was considered important to include the variability of moisture availability (LGP variability), it proved to be highly

*Table 5*

## PAGE Global Agroecosystems by Area and Population

| Global Agroecosystems | Shares Within PAGE Agricultural Extent | | Shares Within Global Total | |
|---|---|---|---|---|
| | Area | Population | Area | Population |
| | *(percentage)* | | | |
| **Temperate** | 38.3 | 35.3 | 11.0 | 26.0 |
| Irrigated and mixed irrigated/rainfed | 4.6 | 13.5 | 1.3 | 10.0 |
| Rainfed, humid/subhumid, flat | 10.8 | 11.3 | 3.1 | 8.4 |
| Rainfed, humid/subhumid, sloping | 6.5 | 6.2 | 1.8 | 4.6 |
| Rainfed, semiarid/arid | 16.5 | 4.2 | 4.7 | 3.1 |
| **Moderate Cool/Cool/Cold Tropics** | 4.1 | 3.1 | 1.2 | 2.3 |
| Irrigated and mixed irrigated/rainfed | 0.3 | 0.5 | 0.1 | 0.3 |
| Rainfed, humid/subhumid | 3.7 | 2.7 | 1.1 | 2.0 |
| **Moderate Cool/Cool/Cold Sub-Tropics** | 19.5 | 20.3 | 5.6 | 14.9 |
| Irrigated and mixed irrigated/rainfed | 3.9 | 6.4 | 1.1 | 4.7 |
| Rainfed, humid/subhumid | 11.0 | 12.1 | 3.2 | 8.9 |
| Rainfed, semiarid/arid | 4.6 | 1.8 | 1.3 | 1.3 |
| **Warm Tropics and Sub-Tropics** | 38.1 | 41.3 | 10.9 | 30.4 |
| Tropics, irrigated and mixed irrigated/rainfed | 2.7 | 8.0 | 0.8 | 5.9 |
| Subtropics, irrigated and mixed irrigated/rainfed | 3.4 | 12.4 | 1.0 | 9.1 |
| Rainfed, humid, flat | 4.7 | 4.8 | 1.3 | 3.5 |
| Rainfed, subhumid, flat | 7.4 | 4.1 | 2.1 | 3.0 |
| Rainfed, humid/subhumid, sloping | 7.6 | 4.7 | 2.2 | 3.4 |
| Rainfed, semiarid/arid, flat | 11.0 | 6.4 | 3.1 | 4.7 |
| Rainfed, semiarid/arid, sloping | 1.3 | 1.0 | 0.4 | 0.7 |
| Total | 100.0 | 100.0 | 28.6 | 73.6 |

**Source:** Global Agroecosystems: IFPRI calculation based on FAO/IIASA 1999; GLCCD 1998; USGS EDC 1999; Döll and Siebert 1999. Population share: IFPRI calculation based on CIESIN 2000.
**Note:** The population shares include both rural and urban populations residing within the global extent of agriculture. If the population for urban areas, as defined by the Stable Lights and Radiance Calibrated Lights of the World database (NOAA-NGDC 1998; Elvidge et al. 1997), are excluded, the percentage of total population living within the extent of agriculture drops to about 50 percent.

and inversely correlated to the length itself: areas with a lower average number of growing season days also tended to have more variation in growing season length. For this reason, LGP variability was not included in the final characterization. Combining the thermal zone, LGP, slope, and irrigation variables yielded the pilot agroecosystem classes depicted in Map 5. We chose the value ranges of the individual variables that define the 16 classes depicted in the map by maintaining a reasonable balance in the share of both land area and population among classes. The spatial population density data came from an inventory of national censuses which were compiled by administrative units and then standardized to 1995 and translated into a global gridded surface at a resolution of approximately 5km by 5km (CIESIN 2000).

Table 5 summarizes both proportionate area and estimated population shares within each aggregate agroecosystem class.

The table suggests that almost three quarters of the world's population live within the extent of agroecosystems as defined by the satellite derived agricultural extent plus additional data on irrigated areas. This finding should be treated with some caution as the population density surface does not distinguish between urban and rural populations,[7] and the satellite data likely underreports extensive agricultural lands. However, the data do emphasize the close linkage between agricultural production potential and the evolution of human settlement. Furthermore, there appears to be a bias toward settlement in the more favorable agricultural areas. Irrigated and mixed irrigated/rainfed agroecosystems occupy some 15 percent of the agricultural extent, but around 40 percent of the populations living within the agricultural extent are located in those agroecosystems. Conversely, while arid and semiarid agroecosystems comprise around one third of the agricultural

Box 1

## Changes in Cultivated Land in China, 1988-95

| Increase | Stock | Decrease |
|---|---|---|

1995  131.1
1994  131.5
1993  132.0
1992  132.3
1991  132.6
1990  132.6
1989  132.5
1988  132.5

Conversion
Re-use
Drainage
Reclamation

Conv. to horticulture
Conv. to forestry land
Loss due to disaster
Conv. to grassland
Conv. to fish ponds
Settlements
Water conservancy
Transp. Infrastructure
Industrial Sites
Public buildings

1000 ha    Mill ha    1000 ha

**Data Source:** State Land Administration, Statistical Information on the Land of China in 1995. Beijing, 1996. Equivalent reports were used for 1988 to 1994.

The chart illustrates how the observed gradual change in China's stock of cultivated land arises from a more dynamic combination of simultaneous land use changes that increase and decrease cultivated land in different locations. In 1995, for instance, China lost some 798.1 thousand hectares of cultivated land. Most of it was converted to horticulture, used for reforestation, or was lost in disasters—mainly floods and droughts. However, China's farmers also expanded the cultivated land by some 388.9 thousand hectares—mainly by reclaiming previously unused areas, but also by converting areas previously used for other purposes. The net change of these increases and declines, which amounted to some 409.1 thousand hectares, reduced the stock of cultivated land only slightly. Some general trends can be inferred from this chart: (1) Approximately 70 to 75 percent of China's cultivated land

losses are not what people usually imagine—a permanent transformation of cropland into infrastructure or urban areas. Most cultivated land losses are conversions into other types of agricultural use or losses because of disasters. Infrastructure, settlements, and industries account for only some 10-15 percent of the losses. (2) There is a clear trend of growing decreases since 1990—a year when the decreases where actually smaller than the increases of cultivated land. This trend is not matched by an equivalent amount of reclamation, which results in a growing net loss of cultivated land in China.

**Source:** CD-ROM: Can China Feed Itself? Heilig, G.K. 1999. See also Fisher, G., Y. Chen, and L. Sun. 1998. http://www.iiasa.ac.at/Research/LUC/ChinaFood/.

extent, they contain only around 13 percent of its population. The data suggest that agricultural land-quality differences related to water availability are much more significant from a human settlement perspective than broad latitudinal differences. Population distribution among the temperate, subtropical and tropical agroecosystems divisions were remarkably close to the agroclimatic area shares. For example, about 38 percent of the agricultural extent lies in temperate regions, as does around 35 percent of the population.

## Global Agricultural Land Uses: Information Status and Needs

We have identified two major shortcomings of land use related data. The first is the lack of disaggregated data on land use conversion that would allow a true picture of land use dynamics and related ecosystem impacts to be assessed. The current reporting of *net* change can give a misleading impression of underlying land use flux and spatial shifts in agricultural expansion, abandonment, and conversion. This weakness could be overcome if satellite image interpretations could more reliably detect major agricultural land cover types and such interpretations were made available on a more routine basis.

*Table 6*

## Characterizing Production Systems

| Attributes | Range of Attribute Values[a] | |
|---|---|---|
| | **Typically associated with less commercialized and low external-input systems** | **Typically associated with highly commercial, high external-input systems** |
| **Land and Resources** | | |
| Tenure Status | Community-based, customary. | Private, titled. |
| Title or Transfer Practices | Inheritance, gift, rent/sharecropping. | By sale/lease. |
| Use Rights | Multiple uses/users. Variable access, withdrawal, and transformation rights. | Individual. Fixed. Legally established. |
| Boundaries | Overlapping, transitory, informal. | Formal or legal demarcation. |
| Plot Size, Distribution | Small plots, fragmented. | Large plots, consolidated. |
| Natural Resources | Common property, collectively managed. | Individually managed and accessed. |
| **Management** | | |
| Objectives | Risk adjusted output maximizing. | Long term profit maximizing. |
| Productivity Strategies | Intensive use of environmental endowments (radiation, natural soil fertility) and internal resources (recycling). Intensive use of labor and draft animal power. | Control environment. Intensive use of capital (physical and human). |
| Cropping Systems | Polyculture (landraces and local cultivars). Mixed cropping often in multiple canopy layers. Asynchronous cultivation. Broad base of species and genetic diversity. | Monoculture (high yielding varieties). Specialization. Synchronous cultivation. Limited species and genetic diversity. More temporal change of germplasm. |
| Nutrient Management | Animal and green manure, fallow, recycling. | Primarily mineral fertilizer. Animal manure and legume rotations. |
| Pest/Weed Management | None or biocontrol/manual weeding. | Pesticides/herbicides. |
| Erosion Management | Mulch, litter, vegetative barriers. | Contour plowing, leveling. |
| Water Management | Manually constructed, gravity fed irrigation. | Irrigation pumps and delivery systems. |
| Livestock Management | Herding, free range. | Stall-fed, feedlots, intensive dairy. |
| **Inputs** | Own labor, retained seed, draft animal power, animal and green manure and crop residues (e.g., recycling). Indigenous knowledge. | Drainage, irrigation, HYV seeds, machinery, fuel, fertilizers, agrochemicals. Scientific knowledge, information intensive. |
| **Outputs** | Locally diverse. Food/shelter subsistence focus. | Uniform quality and traits. Market focus. |
| **Other Key Factors** | Population pressure. Limited access to credit, and likely limited infrastructure and services. | Credit, market accessibility, and integration. |
| **Ecosystem Concerns** | | |
| On-site | Excessive land conversion and related biodiversity, and carbon storage losses. Excessive nutrient mining when intensified. Soil erosion on steeper slopes. | Pesticide: impacts on other fauna, farm worker health; increased resistance. Reduced biodiversity. Salinization in irrigated areas. Over-extraction of water resources. |
| Off-site | Sediment effect from erosion. Increased carbon emissions from land use conversion and burning. | Agrochemical leaching/damage to aquatic systems. Higher water treatment costs. Agrochemical risk to food consumers. Methane emissions from paddy/livestock. Over-extraction of water resources. |

**Note:** (a) No single system is likely to have all the attributes of either column. There are also commercial systems, such as organic farming, that rely on low external-input production practices.

To improve the assessment of agricultural land dynamics and potential ecosystem impacts, additional information is required on conversions to and from agricultural land, as well as between important agricultural land uses within the agricultural extent. For example, Box 1 shows a disaggregated balance of land use conversion that underlay changes in the overall stock of cultivated land in China from 1988 to 1995 (cultivated land stock is shown as the central vertical column). In China, cultivated land, used primarily for the production of annual grains, pulses, and roots and tubers, is distinguished from land used for horticultural crops, so conversion to and from horticulture is reported as a separate land-conversion category. It is clear that looking solely at year-to-year net changes in land use gives an unrealistically conservative impression of the true dynamics of land use change.

The second major weakness is the paucity of data on the production system and resource management aspects of land use. This level of detail shapes the likely trajectory of agroecosystem performance in delivering agricultural and environmental goods and services. Unfortunately, land management can change relatively rapidly, particularly as a response to the ebb and flow of market opportunities, and land use information is difficult to collect. It is not amenable to collection by remote sensing or conventional statistical reporting. However, many countries collect agricultural census and survey data and a growing number now use spatial referencing tools such as global positioning systems (GPS).

The importance of obtaining land use and land management data has long been recognized. The IGBP Land Use Cover and Change (LUCC) taskforce is charged with accelerating the collection and characterization of such data (in addition, and as a complement to, land cover data) (IGBP 2000). At a global scale, several proxy measures of agricultural land use exist, including the following: statistical information on crop types, areas, and yields; animal populations and products; irrigated areas; fertilizer inputs; agricultural labor force; and use of traditional versus modern varieties. Although these all provide some broad notion of land use in the aggregate, they are seldom explicitly linked to each other. Furthermore, they shed little light on the scale of agroecosystem enterprises, the temporal and spatial arrangement of production system components, and other agronomic and management practices that determine the local capacity of each agroecosystem to deliver goods and services in a stable and sustained manner. Table 6 lists a number of attributes that might be used to characterize agroecosystems, as well as some typical attribute values for two polar cases of low and high external-input production systems. But the range, complexity, and time dependent nature of such information generally make it difficult to assemble.

# FOOD, FEED, AND FIBER

Food outputs from the world's agroecosystems have more than kept pace with global population growth and as of 1997 provided, on average, 24 percent more food per person than in 1961, in spite of the 89 percent growth in population that occurred over the same period.[8] Despite the impacts of global economic shocks induced by the oil crises of the mid 1970s and early 1980s, improvements in agricultural productivity have seen food prices drop by around 40 percent in real terms (*see Figure 8*). This represents a significant improvement in human welfare, particularly for poor consumers, who spend a large share of their income on food, as well as for those farming households able to take advantage of new production technologies. But despite these major achievements, the full benefits of more and cheaper food are still beyond the reach of the estimated 790 million of the world's poorest people who are chronically undernourished (FAO 1999a:29).

## Global Production Patterns

An important factor in determining the likely impacts of food, feed, and fiber production on the condition of agroecosystems is the amount of land occupied by different crop and animal production systems. The cultivation of some crops, such as cassava, when poorly managed, can expose soil to water erosion and rapidly exhaust soil fertility. Other systems, such as soybean, if properly managed, can help fix nitrogen in the soil and provide ground protecting cover. Large amounts of pesticide and fertilizers are often applied to potatoes and other vegetables.

*Figure 8*

**Global Index of Food per Capita and Food Prices**

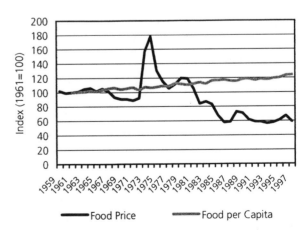

**Source:** Food per Capita: FAOSTAT 1999; Food Prices, 1959–80: IMF 1987; 1980–98: IMF 1998.

*Table 7*

## Regional Distribution of Crops by Area Harvested, 1995–97 Average

| Region | Cereals | Maize | Rice | Wheat | Fiber Crops | Fruits | Oil Crops | Pulses | Roots and Tubers | Sugar Crops | Other | Total | Percentage of Total Crop Harvested Area |
|---|---|---|---|---|---|---|---|---|---|---|---|---|---|
| | | | | | | (Mha) | | | | | | | |
| North America | 82.5 | 29.6 | 1.2 | 36.9 | 0.1 | 1.3 | 40.5 | 2.0 | 0.7 | 1.0 | 2.2 | 130.3 | 11.0 |
| Latin America and the Caribbean | 50.3 | 29.1 | 6.5 | 9.3 | 0.3 | 6.6 | 27.1 | 8.7 | 4.3 | 8.5 | 10.3 | 116.2 | 9.8 |
| Europe | 63.8 | 11.1 | 0.4 | 26.7 | 0.1 | 7.3 | 13.8 | 2.6 | 3.7 | 3.0 | 5.0 | 99.3 | 8.4 |
| Former Soviet Union | 92.1 | 2.7 | 0.5 | 47.6 | 0.3 | 3.0 | 10.5 | 2.7 | 6.3 | 2.5 | 2.3 | 119.8 | 10.1 |
| West Asia/North Africa | 42.5 | 1.9 | 1.4 | 25.5 | 0.0 | 4.2 | 6.3 | 4.1 | 0.8 | 0.9 | 5.3 | 64.1 | 5.4 |
| Sub-Saharan Africa | 80.2 | 24.6 | 6.7 | 2.9 | 0.4 | 7.2 | 22.4 | 13.7 | 17.0 | 1.2 | 12.9 | 154.8 | 13.0 |
| East Asia | 96.9 | 24.4 | 35.3 | 30.1 | 0.4 | 9.4 | 26.5 | 3.5 | 10.3 | 1.9 | 16.7 | 165.6 | 14.0 |
| South Asia | 130.3 | 8.1 | 57.8 | 37.2 | 1.5 | 4.4 | 42.6 | 26.7 | 2.1 | 5.3 | 11.6 | 224.6 | 18.9 |
| Southeast Asia | 49.5 | 8.4 | 40.5 | 0.1 | 0.8 | 3.2 | 16.1 | 3.0 | 4.1 | 2.1 | 12.1 | 90.8 | 7.7 |
| Oceania | 15.9 | 0.1 | 0.1 | 10.2 | 0.0 | 0.4 | 1.7 | 2.1 | 0.3 | 0.5 | 0.4 | 21.2 | 1.8 |
| World | 704.0 | 139.9 | 150.6 | 226.5 | 3.9 | 47.0 | 207.6 | 69.2 | 49.5 | 26.8 | 78.8 | 1,186.8 | 100.0 |

**Source:** Compiled from FAOSTAT 1999.

And, globally, chicken and pig production are becoming increasingly industrial in scale, often giving rise to significant local pollution problems.

At a local level, the structure of production and its evolution over time are closely related to the level of agricultural commercialization. As communities become more actively involved with markets, farm production becomes less focused on self-sufficiency and increasingly on producing those commodities for which the land, the farmer, or the community has some economic comparative advantage. Thus, trading opportunities lead to increasingly divergent patterns of production and consumption at the local level. In such cases, farm households are more interested in the monetary value, rather than the nutritional value of local food and fiber production. For these reasons the food, feed, and fiber indicators presented here include a variety of area, yield, dietary energy, and monetary measures.

## CROP DISTRIBUTION

Some 704 million hectares, almost 60 percent of the world's cropland, is dedicated to the production of cereals. Only in Latin America, with some 43 percent of cereals, does that share fall below half of all harvested land (*see Table 7*). Wheat, rice, and maize are the dominant cereals, occupying 32, 21, and 20 percent of cereal crop area, respectively. The irrigated and mixed humid and subhumid agroecosystems of Asia contain a staggering 89 percent of the world's harvested rice area. Wheat areas are more broadly spread across temperate and subtropical rainfed and irrigated systems, with the Former Soviet Union, South Asia, and North America accounting for over half the

global total. Maize is even more widely distributed partly because of its broader agroecological adaptability and partly because of its ubiquitous use as an animal feed source. Latin America, North America, Sub-Saharan Africa, and East Asia all harvest between 24 and 30 Mha of maize per year. Although on average, about 66 percent of global maize production is used for animal feed, the developed-country average is around 76 percent and the developing around 56 percent (CIMMYT 1999:62).

The 208 Mha of oil crops, of which soybean occupies some 67 Mha, represents the second most common crop group. In addition to their oil content, these are important sources of high-quality protein animal feed. About 80 percent of processed soybean (the cake left following the removal of oil for human consumption) can be used as animal feed. Cereals and oil crops occupy about 77 percent of the area of the world's croplands. The remainder of cropland is occupied by pulses (5.8 percent), roots and tubers (4.1 percent), fruits (4.0 percent), sugar crops (2.2 percent), and other (6.5 percent).

Agriculture produces more than food. Crops producing fiber (such as cotton, flax, sisal), medicines, dyes, chemicals, and other non food industrial raw materials account for nearly 7 percent of harvested crop area. Some food crops are grown primarily for fuel (such as ethanol) but more woody biofuels are being planted. Currently available agricultural statistics exclude most of the world's production of trees, shrubs, and palms grown on farms for woodfuel, construction material, and raw materials for household artesanry or local processing.

*Figure 9*

## Distribution of Ruminant Livestock by Production Systems[a,b] and Agroclimatic Regions[c], 1992-94 average

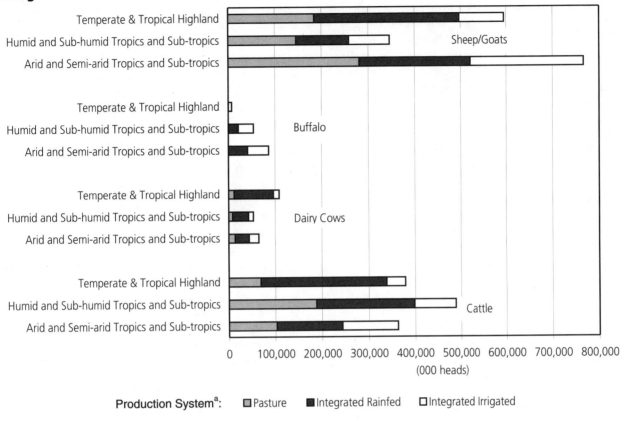

Production System[a]:  ☐ Pasture   ■ Integrated Rainfed   ☐ Integrated Irrigated

**Source:** IFPRI calculation based on FAO world livestock production systems (Seré and Steinfeld 1996).

**Notes:** (a) Pasture—more than 90 percent of feed comes from grassland, forages, and purchased feeds; Integrated Rainfed—more than 10 percent of the dry matter feed comes from crops, by-products, or stubble; Integrated Irrigated—same as Integrated Rainfed plus more than 10 percent of the value of non-livestock farm production is from irrigated land. (b) excludes confined ("landless") animal production estimated at 66,231 thousand heads for cattle and 9,931 thousand heads for sheep and goats. (c) Agroclimatic regions were based on FAO's agroecological zones project (FAO 1982).

## LIVESTOCK DISTRIBUTION

The global ruminant-livestock population includes approximately 1,225 million head of beef cattle, 227 million head of dairy cattle, 148 million head of buffalo, and 1,708 million head of sheep and goats. Latin America supports the largest beef cattle population of approximately 350 million head (26 percent), and West Asia/North Africa (WANA) the smallest with 8 million head (under 1 percent), while over 96 percent of the world's buffaloes are in Asia, which also has the largest share of the sheep and goat population (30 percent). By production system (*see Figure 9*), cattle are fairly uniformly spread across major agroclimatic zones, with a slightly higher share in the tropical and subtropical humid and subhumid areas (39 percent). Dairy cattle are predominantly found in temperate and highland ar-

eas (48 percent), while sheep and goats predominate in tropical and subtropical arid and semiarid ecosystems (45 percent). Integrated rainfed crop-livestock systems dominate in the production of beef cattle (50 percent), dairy cattle (66 percent), and sheep and goats (39 percent). However, for Sub-Saharan Africa and Latin America, pastures are dominant for cattle (67 percent and 50 percent of animals respectively) (Seré and Steinfeld 1996).

Over the past 30 years, livestock production has approximately tripled compared to a doubling of crop output (Pinstrup-Andersen et al. 1999). The most recent surge in demand for meat and milk, particularly in developing countries, has given rise to what has been dubbed the "Livestock Revolution" (*see Box 2*).

*Box 2*

## Livestock Revolution

There is increasing evidence of a substantial livestock revolution taking place, particularly in the developing world. Unlike the Green Revolution, this change is driven by significant increases in demand for meat products—particularly poultry and pig meat—and milk, which in turn are translating into growing demand for feed grains and high-protein meal. This rapid growth should offer greater income opportunities and improved nutrition, particularly to the poor, but will likely pose additional environmental and human health problems.

Over the period 1982-94, the global demand for meat grew by 2.9 percent per year, but this change comprised a growth rate of 1 percent for developed countries and 5.4 percent for developing countries. Within some specific meat categories, developing-country growth rates were even higher—pork 6.2 percent per year, poultry 7.6 percent per year, and milk at 3.1 percent per year. By 1993, there were about 878 million head of pigs and some 13 billion chickens globally. About 36 percent of the pig population was in the developed countries and of the 64 percent in the developing world, 44 percent was in China alone with the next largest region, Latin America, having 9 percent. In the case of chicken, the overall developed/developing-country share of 35/65 percent is spread much more evenly across regions.

Both pig and poultry operations in developing countries are likely following the same type of intensive landless production prevalent in developed countries and are directly importing the necessary technologies to operate industrial-scale pig and poultry operations near urban centers. As we show in the box on *Cheaper Chicken, More Pollution*, even in the United States where environmental regulation enforcement powers are established and applied, such concentrated industrial-scale installations can lead to significant local pollution and health hazards. Other health risks, particularly in developing countries, could arise through animal-borne diseases, such as avian flu and salmonella, microbial contamina-

tion from unsafe handling of foods, and a build-up of pesticides and antibiotics in the food chain through production practices.

The growth in meat demand also implies an additional demand for feed cereals (especially maize), oil crop cake, and other crop residues, such as those from cassava and sugar processing. Analysis by the International Food Policy Research Institute (IFPRI) suggests that under pessimistic future scenarios this demand would increase real maize prices by around 20 percent by the year 2020, to about the same price levels that prevailed in the early 1980s.

From the perspective of the poor, the livestock revolution offers many opportunities. The rural poor and the landless, especially women, could get a higher share of their income from livestock than better-off rural people (with the main exceptions found in areas with large-scale ranching, such as parts of Latin America). Furthermore, livestock provides the poor with fertilizer, draft power from cattle and buffalo, and in some areas fuel, along with other opportunities to diversify income. Livestock products also benefit the poor by alleviating the protein and micronutrient deficiencies prevalent in developing countries. Increased consumption of even small additional amounts of meat and milk can provide the same level of nutrients, protein, and calories to the poor that a large and diverse amount of vegetables and cereals could provide. Such synergies could also promote the broader scale adoption of integrated crop-livestock systems, which with proper management can be highly sustainable production systems for smallholders in developing countries.

*Source*: Adapted from Delgado, C., M. Rosegrant, H. Steinfeld, S. Ehui, C. Courbois. "Livestock to 2020: The Next Food Revolution." Food, Agriculture, and the Environment Division Discussion Paper No. 28. Washington, D.C.: IFPRI.

## Input Intensity of Agricultural Production Systems

Although there are a relatively limited number of globally important agricultural commodities—some 30 commodities supply more than 90 percent of the world's calorie consumption—the ways in which those commodities are produced varies significantly (FAO 1998:14). Furthermore, the specific mix of inputs and production technology applied has a direct bearing on the long-term capacity of agroecosystems to deliver agricultural and environmental goods and services. Production systems can be characterized by the extent to which they rely on the following: polyculture versus monoculture; traditional versus modern

varieties; inorganic fertilizers versus recycling of crop residues and manures; pesticides versus biotic control; rainfall versus the supplemental application of water; and hand labor versus machinery. Seeds and water are considered separately in subsequent sections; so here we focus on indicators of labor, fertilizer, machinery, and pesticides (summarized at a regional level in Table 8).

### AGRICULTURAL LABOR

The labor input-intensity measure is defined as agricultural labor per hectare of cropland.[9] Across all global agroecosystems the average labor utilization is 0.85 person per hectare (pph), but varies widely according to labor scarcity (wage rates), pro-

*Table 8*
## Input Intensity Indicators, 1995–97 Average

| Region | Agricultural Labor | Tractors[b] | Inorganic Fertilizer[a] | | | | Irrigated Share of Cropland |
|---|---|---|---|---|---|---|---|
| | | | N | P₂O₅ | K₂O | Total | |
| | (person per hectare) | (hectare per tractor) | (kilogram per hectare) | | | | (percent) |
| North America | 0.02 | 41 | 57.1 | 21.6 | 23.1 | 101.8 | 9.8 |
| Latin America and the Caribbean | 0.28 | 102 | 26.7 | 18.3 | 17.1 | 62.1 | 11.3 |
| Europe | 0.15 | 14 | 89.7 | 32.2 | 36.5 | 158.4 | 12.5 |
| Former Soviet Union | 0.11 | 102 | 14.0 | 4.5 | 2.3 | 20.8 | 9.3 |
| West Asia/North Africa | 0.45 | 60 | 39.7 | 18.1 | 3.3 | 61.1 | 26.4 |
| Sub-Saharan Africa | 0.98 | 622 | 6.1 | 3.4 | 2.1 | 11.6 | 3.7 |
| East Asia | 3.58 | 47 | 130.7 | 51.1 | 83.2 | 265.0 | 38.7 |
| South Asia | 1.57 | 123 | 62.9 | 19.3 | 6.6 | 88.8 | 38.0 |
| Southeast Asia | 1.47 | 232 | 50.2 | 16.6 | 17.0 | 83.8 | 17.4 |
| Oceania | 0.05 | 138 | 17.7 | 25.5 | 6.8 | 50.0 | 5.2 |
| World | 0.85 | 57 | 53.2 | 21.0 | 15.5 | 89.7 | 17.5 |

**Source:** Compiled from FAOSTAT 1999.
**Notes:** Labor, fertilizer, pesticide, and tractor inputs are expressed based on hectares of cropland (annual plus permanent crops). (a) Includes only commercial inorganic fertilizers: Nitrogen (N), phosphorus (P₂O₅) and potassium (K₂O). (b) Tractors are defined here as all wheel and crawler tractors (excluding garden tractors) used in agriculture.

duction structure, production technologies, energy and machinery costs, and so on. Thus, in land-abundant, high-income countries with extensive animal production, such as the United States and Australia, the average labor input intensity across all agroecosystems is around 0.01 pph and 0.02 pph respectively. In (agricultural) land scarce, low-income regions having irrigated and mixed irrigated/rainfed crop-based agroecosystems, such as East Asia, labor inputs average about 3.6 pph. Across the Asian agroecosystems the average labor input always exceeds 1.4 pph *(see Map 6 and Table 8)*.

## TRACTORS

One indicator of the nature of production systems is the extent to which agricultural machinery, specifically tractors, are used. Tractors offer labor-saving increases in productivity where labor is scarce (relatively costly) and perform a range of traction and transport functions that might be uneconomic manually. They are, however, expensive to purchase and maintain, are more suited to flatter lands, and their overuse can lead to soil-management problems, such as compaction and other changes in physical properties.

As a global average there are about 57 hectares of cropland per tractor. In the highly commercialized agriculture of Europe and North America, where high labor costs prevail, there are about 14 and 41 hectares per tractor, respectively. In Asia, characterized by lower incomes, high population density, and much smaller land holdings, there are approximately 47, 123, and 232 hectares per tractor respectively in East Asia, South Asia,

and Southeast Asia *(see Map 7 and Table 8)*. Sub-Saharan Africa averages around 600 hectare per tractor. These averages need to be treated with caution as there is no differentiation by tractor type or quality—for example, by horse-power rating.

## INORGANIC FERTILIZERS

An important input-intensity indicator is the extent to which inorganic (mineral) fertilizers are applied. In many agroecosystems, nutrients extracted by crops and pastures are replenished, to varying degrees, by the application of inorganic fertilizers containing nitrogen (N), phosphorus (P₂O₅), and potassium (K₂O). Too little replenishment, from inorganic or organic sources, leads to long-term nutrient depletion that can exhaust the inherent soil fertility. Excess or mistimed fertilizer application can cause nutrient runoff or leaching and consequent soil and water pollution problems. The most commonly observed consequences of such pollution are the eutrophication of water bodies, and the damage to other forms of aquatic life and downstream water uses caused by algal blooms.

Global fertilizer consumption stands at about 128 million tons per year (1997) and has been in general decline since 1989, although the surge in cereal production from 1995 to 1997 reversed this downward trend, at least temporarily (FAOSTAT 1999). It is estimated that about 55 percent of fertilizers are applied to cereals, 12 percent to oil crops, 11 percent to pasture and hay, about 6 percent to roots and tubers, and about 5 percent to fruit and vegetables (Harris 1998:19). Nutrient application rates are important indicators of both the potential for

*Table 9*
## World Pesticide Consumption, 1983–98

| Region | Value[a] 1983 | Value[a] 1993 | Value[a] 1998 | 1998 Value per ha of Cropland[b] | Share 1998 | Compound Growth Rate[a] 1983-93 | Compound Growth Rate[a] 1993-98 | 1992 Pesticide Share[a] Herbicides | 1992 Pesticide Share[a] Insecticides | 1992 Pesticide Share[a] Fungicides | 1992 Pesticide Share[a] Other |
|---|---|---|---|---|---|---|---|---|---|---|---|
| | *(US$ millions)* | | | *($ per ha)* | *(percent)* | *(percent per year)* | | | | | |
| North America | 3,991 | 7,377 | 8,980 | 40 | 26 | 6.3 | 4 | 66 | 22 | 8 | 5 |
| Latin America | 1,258 | 2,307 | 3,000 | 19 | 9 | 6.3 | 5.4 | 47 | 29 | 19 | 4 |
| Western Europe | 5,847 | 7,173 | 9,000 | 102 | 26 | 2.1 | 4.6 | 43 | 18 | 30 | 9 |
| Eastern Europe | 2,898 | 2,571 | 3,190 | 14 | 9 | -1.2 | 4.4 | 38 | 39 | 18 | 5 |
| Africa/Mideast | 942 | 1,258 | 1,610 | 5 | 5 | 2.9 | 5.1 | | | | |
| Asia/Oceania | 5,572 | 6,814 | 8,370 | 16 | 25 | 3 | 4.4 | | | | |
| Japan | | 3,545 | | | | | | *31* | *34* | *33* | *2* |
| Far East | | 2,600 | | | | | | *31* | *48* | *14* | *7* |
| Total | 20,507 | 27,500 | 34,150 | 23 | 100 | 3 | 4.4 | 45 | 29 | 19 | 6 |

**Source:** (a) Yudelman et al. 1998:10. (b) IFPRI calculation based on Yudelman et al. 1998:10 and FAOSTAT 1999.

yield enhancement as well as for nutrient mining or leaching. A recent global survey of fertilizer usage found that vegetables as a group have the highest fertilization rates, some 242 kilograms per hectare (kg/ha), followed by sugar crops (216 kg/ha), and roots and tubers (212 kg/ha). Cereals and oilseed, the dominant crops by area and by the total volume of applied fertilizer, receive about 102 kg/ha and 85 kg/ha respectively (IFDC/IFA/FAO 1997 cited in Harris 1998:5).

At the individual crop level, the average application rate of fertilizers for banana, at 479 kg/ha, far outstrips the next most fertilized crops: sugar beet (254 kg/ha), citrus (252 kg/ha), potato (243 kg/ha), vegetables (242 kg/ha), and palm oil (242 kg/ha). The fertilization practices of such crops clearly have potential to cause pollution damage. Among cereals, maize is the most fertilized at 136 kg/ha, while wheat and rice receive 116 kg/ha and 112 kg/ha, respectively. However, there are wide variations. The average fertilizer application rate for wheat varies from just under 20 kg/ha in Myanmar and Nepal to more than 300 kg/ha in Japan (IFDC/IFA/FAO 1997 cited in Harris 1998:5-8).

As shown in Table 8 and Map 8, aggregate fertilizer use (average NPK applied per hectare of cropland)—the selected fertilizer intensity indicator—varies from around 265 kg/ha in East Asia, through 102 and 158 kg/ha applied in North America and Europe respectively, to a low of around 12 kg/ha in Sub-Saharan Africa (FAOSTAT 1999).

Although increasing nutrient supply is recognized as an essential step in raising agricultural productivity in Sub-Saharan Africa, there is much debate about the most appropriate strategies for achieving this. For both technical and economic reasons, the use of inorganic fertilizer alone is generally seen as inappropriate. Technical reasons include the need for enhanced soil organic matter content as a precondition for the effective-

ness of inorganic fertilizers. Economically, the credit institutions and fertilizer markets accessible to poor producers are often absent or not well developed. Preferred nutrient management approaches include increased use of organic fertilizers such as green and animal manures and crop residues, and other practices that maintain and improve not only fertility but also water-holding capacity and organic-matter content. To achieve the productivity levels necessary to meet the rapidly growing food demands of Sub-Saharan Africa will most likely require a hybrid form of nutrient management, emphasizing strategic use of both organic and mineral fertilizers.

Fertilizer input alone is not a sufficient indicator of the long-term productive capacity of soil. Over time, it is the balance of net nutrient inputs and outflows through crop harvesting, soil erosion, leaching and so on, that are more important. Soil nutrient balances are considered further in the Soil Resource Condition section.

## PESTICIDES

A central strategy in improving agricultural output is to limit losses from the effects of pests and diseases and from weed competition. Since the mid 1900s, the approach to crop protection has relied increasingly on the use of pesticides (defined here to include insecticides, nematocides, fungicides, and herbicides). Data on pesticide use at a regional and global level is limited. Data on its intended and unintended impacts are even more scarce and often controversial.

Global estimates suggest that while current losses in wheat through the impacts of pests, diseases, and weeds are around 33 percent they might rise to 52 percent without the use of control measures. Maize losses could also potentially increase by approximately 20 percent, from 39 to 60 percent, and rice losses could increase 30 percent, from 52 to 83 percent (Oerke et al.

1994 cited in Gregory et al. 1999:235). If such estimates are even broadly correct, the negative impacts of reduced pest control effectiveness on farmer incomes and consumer prices would be extremely significant. Certainly, pesticide use has risen and continues to increase dramatically *(see Table 9)*, indicating that farmers find pesticides cost-effective from a production perspective, particularly where alternative forms of crop protection are labor intensive and labor costs are high.

Other economic and social dimensions of pesticide use cause widespread environmental and human health concerns and have raised questions about its scientific rationale. Environmental concerns focus on the toxicity of many pesticides to biological species other than those directly targeted, such as soil microorganisms, insects, plants, fish, mammals, and birds, that might not only be beneficial to agriculture or other human economic activities, but that are part of a biodiversity valued by society for recreational, cultural, ethical or other reasons. Although regulations in many countries have promoted the development of more-specific, less toxic, and more rapidly decomposing pesticides, many of the more damaging pesticides are still marketed in countries where regulations are more lax. Human health concerns include the effects of ingesting chemical residues contained in foods, but also the ill-effects on farm workers of pesticide handling and application, particularly in countries where safety standards are not well established, understood, or enforced (Antle and Pingali 1994; Crissman et al. 1994).

Part of the scientific debate on crop protection relates to the ability of pests, weeds, and viruses to develop pesticide resistance. This results in a constant need to develop new pesticide products (or pest resistant plant varieties) to keep one step ahead of biological adaptation. This cycle, dubbed the "pesticide treadmill," has led to biological adaptations resistant to most commercially available pesticides. One global estimate suggests that around 1,000 major agricultural pests (insects, mites, plants, and seeds) are now immune to pesticides (Brown et al. 1999:124), including some 394 insects and mites, 71 weeds, and 160 plant species in the United States alone (The Heinz Center 1999:26). Because of such problems, Yudelman (1998:13) estimates, despite a 10-fold increase in both the amount and toxicity of pesticides in the United States between the 1940s and 1990s, that the share of crops lost to pests actually rose from 30 to 37 percent. Other research has found that economic levels of pesticide application are often much lower than those adopted by farmers. Because pesticide costs are often a relatively small share of total production costs, farmers prefer to overapply pesticide than risk severe crop losses.

Pesticides, however, are not the only means to counter crop loss from pests and diseases. Integrated pest management (IPM) techniques, for example, apply ecological science to enable biological suppression that can often keep pest populations below damaging levels, save pesticide and application costs, and in some cases enhance yields. In the Yunnan Province in southern China, IPM techniques for controlling *rice blast* have been rapidly adopted. In 2000, 42,500 hectares of rice fields were being grown using IPM, while 10 other provinces are reportedly experimenting with the approach (Mew 2000). In Vietnam, IPM techniques are reportedly applied by 92 percent of the Mekong Delta's 2.3 million farm households, and insecticide applications have fallen from an average of 3.4 per farmer per season to just one (IRRI 2000). In Indonesia, a survey of 2,000 farmers trained in and applying IPM techniques found that rice yields had increased by an average of 0.5 tons per hectare and the number of pesticide applications had fallen from 2.9 to 1.1 per season. Furthermore, rice fields under IPM were being recolonized by plant and animal species previously suppressed by pesticide use (van de Fliert 1993; reported in Reijntjes et al. 1999).

IPM is not just a developing country, small scale approach. In the United States there have been considerable advances during the 1990s in developing and promoting IPM methods. This has resulted in significantly positive trends in IPM adoption by large-scale, commercial producers, particularly of cotton and potatoes (Fernandez-Cornejo and Jans 1999).

## Status and Trends in Yields and Production Capacity

An important measure of the current capacity of agroecosystems to produce food is the quantity of crop or animal output they can produce per unit of land or per animal—their yield. To some degree, the agroecological characteristics of the domain in which an agroecosystem is located determine this capacity. For example, the biomass production potential of the warm humid tropics is considerably greater than that of the dry semiarid subtropics. But the potential for crop loss through pest and disease is also greater in the tropics, and higher temperatures and rainfall tend to leach and otherwise deplete soil nutrients more rapidly. Within the broad production potential boundaries defined by such agroecological conditions, the specific production practices and inputs applied determine yield levels.

As populations and markets have grown and agricultural expansion possibilities have declined, it has been necessary to intensify production. Many of the major scientific breakthroughs in developing high yielding crop varieties that respond to more intensive cultivation took place during the last two thirds of the twentieth century *(see Figure 10)*. The intensification process has kept aggregate production growing faster than population and, thus, has increased the per capita availability of food over time. But there is also substantial evidence that intensification pressures in both low- and high-external input production systems can bring about significant environmental problems: soil erosion from hillside production, soil salinization in irrigated

*Figure 10*

## Growth in Wheat Yields

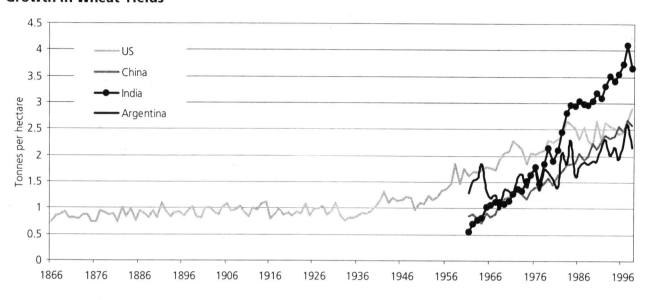

**Source:** Compiled from FAOSTAT 1999; USDA-NASS 1999.

areas, and water pollution from nutrient and pesticide residues, among others. Although slowing down intensification might reduce such environmental externalities on existing agricultural land, proportionally greater amounts of land would be required to meet existing and future food needs. Strategic options to address this problem include policies to foster moderation of population growth rates, more agricultural research, and the development of environmental conservation practices.

The focus of much current agricultural research and development effort is on improving the productivity of agriculture in ways that are more environmentally benign. These so-called "win-win" options should provide simultaneous improvements in both agricultural and environmental goods and services.

At present, the highest yields for practically all commodities that are not exclusively tropical and subtropical are attained either in North America or Europe *(see Table 10)*. Of note, however, are the high average rice yield levels in the irrigated agroecosystems of East Asia, primarily China, at 6.2 tons per hectare. These are only slightly below the highest regional average of 6.6 tons per hectare for North America, but in East Asia these irrigated rice ecosystems occupy approximately 35 million hectares compared to only 1.2 million in North America. Almost without exception the lowest average regional yield levels are found in Sub-Saharan Africa for both crops and livestock. Yields are about average for livestock products for Latin America and the Caribbean (LAC), despite the dominance of the livestock sector, possibly reflecting the much greater emphasis on extensive grazing systems in that region. There are also stark contrasts between the milk yields of Oceania, Europe, and North America, ranging from 2,200 to 7,200 kg per

animal per year, and those of Sub-Saharan Africa and Southeast Asia, at 152 and 163 kg per animal per year, respectively. This regional variability results from differences in climate, herd genetics, levels of specialization, and investments, among others.

The growth rate of cereal yields declined between 1975-85 and 1985-95 in 7 regions that account for approximately 80 percent of global production[10] *(see Table 11)*. Globally, cereal area has been in gradual decline (but less than yield increases, so production has grown), and both area and yields of oil crops have seen continued growth in both periods. Of particular note is the sustained production growth in Sub-Saharan Africa of roots and tubers that globally show little to no growth over the past decade. Analyzing changing production in this way provides useful insights into the relative emphasis of area expansion versus intensification of production and, hence, on the likely nature of environmental consequences of change.

One current concern is that the recently observed slowdown in the growth of cereal yields (Pingali and Heisey 1999) will compromise agriculture's capacity to feed the additional 1.5 billion people expected over the next 20 years, especially since per capita grain consumption also continues to rise. Some fear that this situation reflects stagnation in scientific progress, natural resource degradation, and growing pest resistance. But there is evidence that market factors also drive the trends. Declining commodity prices have caused farmers to make yield-reducing adjustments in input use. Increased attention has been paid to quality enhancement, protein content, grain size and shape, taste, processing qualities, sometimes at the expense of yield. Rising profitability of crops such as soybean and canola has led

Table 10
**Yields of Selected Commodities by Region, 1995–97 Average**

| | North America | Latin America and the Caribbean | Europe | Former Soviet Union | West Asia/ North Africa | Sub-Saharan Africa | East Asia | South Asia | Southeast Asia | Oceania | World |
|---|---|---|---|---|---|---|---|---|---|---|---|
| Yields of Crop Products | | | | | | (Mt/Ha) | | | | | |
| Cereals | 4.5 | 2.6 | 4.5 | 1.4 | 2.0 | 1.1 | 4.8 | 2.2 | 3.1 | 2.0 | 2.9 |
| Maize | 7.6 | 2.5 | 5.8 | 2.9 | 4.4 | 1.4 | 4.8 | 1.6 | 2.3 | 6.5 | 4.0 |
| Wheat | 2.4 | 2.4 | 4.7 | 1.5 | 1.9 | 1.6 | 3.8 | 2.4 | 0.9 | 1.9 | 2.6 |
| Rice | 6.6 | 3.2 | 6.1 | 2.3 | 5.8 | 1.6 | 6.2 | 2.8 | 3.3 | 7.4 | 3.8 |
| Sorghum | 4.1 | 2.8 | 4.9 | 1.0 | 1.6 | 0.8 | 3.9 | 0.8 | 1.5 | 2.1 | 1.4 |
| Millet | 1.5 | 1.2 | 1.5 | 0.7 | 0.8 | 0.6 | 2.0 | 0.8 | 0.7 | 1.1 | 0.7 |
| Potatoes | 35.8 | 13.7 | 22.9 | 11.3 | 19.7 | 8.5 | 14.3 | 15.3 | 13.1 | 33.4 | 16.0 |
| Cassava | – | 12.0 | – | – | – | 8.2 | 15.5 | 21.8 | 12.1 | 11.1 | 9.9 |
| Dry Beans | 1.8 | 0.6 | 0.9 | 1.4 | 1.3 | 0.7 | 1.1 | 0.4 | 0.9 | 0.7 | 0.7 |
| Soybeans | 2.5 | 2.2 | 2.9 | 0.6 | 1.8 | 0.8 | 1.7 | 1.0 | 1.2 | 1.8 | 2.1 |
| Groundnuts | 2.8 | 1.7 | 1.2 | 1.7 | 2.4 | 0.8 | 2.7 | 1.1 | 1.3 | 1.7 | 1.3 |
| Sugar Beets | 45.4 | 65.2 | 48.7 | 17.9 | 34.7 | – | 26.7 | 26.3 | – | – | 35.1 |
| Sugar Cane | 75.4 | 62.7 | 62.7 | – | 99.6 | 54.1 | 63.4 | 63.1 | 60.6 | 86.9 | 63.2 |
| Seed Cotton | 1.8 | 1.4 | 2.8 | 2.0 | 2.7 | 0.9 | 2.8 | 1.0 | 0.7 | 3.4 | 1.6 |
| Yields of Animal Products | | | | | | (Kg/Animal) | | | | | |
| Beef and Veal | 308 | 192 | 248 | 162 | 143 | 139 | 166 | 106 | 169 | 208 | 199 |
| Milk | 7,158 | 967 | 2,185 | 2,055 | 238 | 152 | 543 | 630 | 163 | 3,644 | 931 |
| Pigmeat | 83 | 71 | 84 | 76 | 71 | 46 | 76 | 35 | 59 | 59 | 76 |

**Source:** Compiled from FAOSTAT 1999.

Table 11
**Crop Area and Yield Trends**

| | Cereals | | | | Oil Crops | | | | Roots and Tubers | | | |
|---|---|---|---|---|---|---|---|---|---|---|---|---|
| | 75–85 | | 85–95 | | 75–85 | | 85–95 | | 75–85 | | 85–95 | |
| Region | Area | Yield | Area | Yield | Area | Yield | Area | Yield | Area | Yield | Area | Yield |
| | | | | | (percentage annual growth rate) | | | | | | | |
| North America | 0.25 | 1.77 | -1.02 | 1.20 | 0.55 | 1.27 | 1.69 | 0.87 | 0.21 | 1.30 | 0.81 | 1.29 |
| Latin America and the Caribbean | 0.17 | 3.01 | -1.05 | 2.26 | 1.84 | 2.87 | 0.27 | 3.37 | -0.48 | 0.73 | -0.30 | 0.70 |
| Europe | -0.20 | 3.00 | -1.15 | 0.33 | 5.54 | -0.43 | 1.84 | 0.05 | -2.47 | 0.86 | -3.13 | -0.28 |
| Former Soviet Union | -1.05 | 0.87 | -1.47 | -0.96 | 0.06 | 0.20 | 0.13 | -1.14 | -0.93 | -0.25 | 0.09 | -0.92 |
| East Asia | -0.98 | 4.73 | -0.10 | 1.89 | 2.33 | 3.37 | 1.21 | 1.79 | -1.05 | 1.18 | 0.66 | 1.11 |
| South Asia | 0.39 | 2.38 | -0.24 | 3.06 | 3.00 | 3.80 | 2.03 | 4.30 | 0.67 | 1.78 | 1.11 | 1.76 |
| Southeast Asia | 1.17 | 3.74 | 0.71 | 1.78 | 0.80 | 0.31 | 2.75 | 0.03 | 0.78 | 1.04 | 0.24 | 0.08 |
| West Asia/North Africa | 0.22 | 1.13 | 0.74 | 1.87 | 0.99 | 1.21 | 0.87 | 0.52 | 2.74 | 2.25 | 1.65 | 1.56 |
| Sub-Saharan Africa | 0.52 | 0.26 | 2.49 | -0.65 | 1.87 | 2.38 | 3.04 | 3.05 | 2.35 | 1.25 | 4.02 | 1.71 |
| Oceania | 3.50 | 0.91 | -1.92 | 1.50 | 2.68 | 0.71 | 1.34 | 0.16 | 0.88 | 0.91 | 0.40 | 1.18 |
| World | -0.15 | 1.84 | -0.29 | 1.16 | 1.67 | 2.17 | 1.71 | 1.83 | 0.10 | 0.47 | 1.13 | 0.02 |

**Source:** IFPRI calculation based on FAOSTAT 1999.

Table 12

## Value of Agricultural Production by Region, 1995–97 Average

| Region | Cereals | Roots and Tubers | Fibers | Fruits | Oil Crops | Pulses | Sugar | Livestock | Other | Total | Share of World Value |
|---|---|---|---|---|---|---|---|---|---|---|---|
| | | | | | *(millions of 1989-91 dollars)* | | | | | | *(percent)* |
| North America | 47,880 | 2,829 | 5,483 | 8,807 | 19,782 | 1,461 | 1,370 | 86,611 | 11,180 | 185,404 | 14.0 |
| Latin America and the Caribbean | 17,874 | 4,209 | 1,837 | 16,983 | 12,542 | 3,068 | 9,066 | 63,167 | 11,823 | 140,570 | 10.6 |
| Europe | 37,480 | 9,263 | 784 | 21,363 | 9,980 | 1,644 | 5,043 | 117,386 | 18,194 | 221,136 | 16.7 |
| Former Soviet Union | 17,532 | 7,922 | 2,407 | 3,789 | 1,975 | 835 | 1,513 | 42,045 | 5,086 | 83,103 | 6.3 |
| West Asia/ North Africa | 11,653 | 1,730 | 2,232 | 10,738 | 2,405 | 1,955 | 1,111 | 21,812 | 16,242 | 69,878 | 5.3 |
| Sub-Saharan Africa | 11,891 | 11,762 | 1,788 | 6,367 | 5,003 | 2,529 | 1,065 | 19,640 | 8,720 | 68,767 | 5.2 |
| East Asia | 74,368 | 15,444 | 7,011 | 14,034 | 12,876 | 1,738 | 1,924 | 110,353 | 61,564 | 299,313 | 22.6 |
| South Asia | 48,142 | 3,295 | 6,281 | 9,030 | 10,601 | 9,964 | 5,635 | 40,863 | 20,243 | 154,055 | 11.6 |
| Southeast Asia | 28,097 | 3,187 | 260 | 5,453 | 7,691 | 1,410 | 2,124 | 15,183 | 9,664 | 73,070 | 5.5 |
| Oceania | 4,341 | 366 | 589 | 1,150 | 564 | 689 | 672 | 17,971 | 924 | 27,266 | 2.1 |
| World | 299,259 | 60,007 | 28,673 | 97,713 | 83,419 | 25,294 | 29,524 | 535,031 | 163,641 | 1,322,561 | 100.0 |

**Source:** IFPRI calculation based on FAOSTAT 1999 and FAO 1997a.
**Note:** The total value of agricultural production was formed by weighting 134 primary crop and 23 primary livestock commodities quantities by their respective international agricultural prices for the period 1989-91.

to some displacement of cereals from the most productive farmlands. And economic problems have depressed yields in the Former Soviet Union, a major cereal producing region.

## The Value of Outputs

Agriculture represents around 2, 9, and 23 percent of the total gross domestic product (GDP) of high, middle, and low-income countries respectively. In many of the poorest countries, it often represents 40 to 60 percent of total GDP.[11] Although agriculture represents a mere 1-3 percent of national GDP in the temperate, subhumid, and humid high-income countries of Western Europe and North America, the agricultural GDP of these countries represents about 78 percent of the global total (World Bank 2000:188). Furthermore, there is much evidence that the conventional measure of agriculture's share of total GDP significantly underestimates its true contribution. For example, while the Philippines, Argentina, and the United States were recorded as having agricultural GDP's that comprise 21, 11, and 1 percent of total GDP, respectively, studies have shown that the total "agribusiness" value of agriculture including manufacturing and post-harvesting services comprises 71, 39, and 14 percent of the total GDP in those countries (Bathrick 1998:10).

## VALUE OF AGRICULTURAL PRODUCTION

The global and regional value of agricultural outputs are summarized in Table 12 by major commodity group. The total value of output from the world's agroecosystems is $1.3 trillion per year (IFPRI calculation based on commodity prices in 1989-91 international dollars (FAO 1997a), and average production from 1995-97 (FAOSTAT 1999)). Crops represent 60 percent and livestock 40 percent of this value. Within crops, cereals dominate (38 percent), and together with fruits (12 percent) and oil crops (11 percent) provide over 60 percent of total crop value.

## VALUE PER HECTARE

A comparable indicator of the monetary value of agroecosystem production is the value of output per unit of land.[12] However, the output value is sensitive to the choice of land variable: agricultural land, cropland, or harvested area *(see Table 13 and Map 9)*. Globally, the value of crops per hectare of total cropland is $521 per hectare, while expressing the same total value per unit of harvested land yield is $662 per hectare. Areas with the highest shares of irrigated land, and hence higher cropping intensities, show higher values per unit area of cropland. By all these measures, the intensely cultivated rainfed and irrigated systems of East Asia are most productive. The clear distinction between potential returns to investment in crops versus extensive livestock is revealed in the value of livestock per unit of pasture land. The extensive livestock systems of Sub-Saharan Africa and Oceania exhibit low values per hectare of pasture,

Table 13
## Value of Agricultural Production per Hectare, 1995–97 Average

| Region | Total Value[a] of Output per Hectare of | | Value of Crops per Hectare of | | Value of Livestock per Hectare of Pasture |
|---|---|---|---|---|---|
| | Agricultural Land[b] | Cropland | Cropland | Harvested Area | |
| | *(1989-91 dollars per hectare)* | | | | |
| North America | 370 | 824 | 439 | 756 | 328 |
| Latin America and the Caribbean | 180 | 878 | 482 | 663 | 103 |
| Europe | 1,026 | 1,636 | 766 | 1,042 | 1,462 |
| Former Soviet Union | 145 | 375 | 185 | 341 | 123 |
| West Asia/North Africa | 193 | 744 | 511 | 747 | 81 |
| Sub-Saharan Africa | 69 | 400 | 285 | 316 | 24 |
| East Asia | 438 | 2,067 | 1,304 | 1,139 | 203 |
| South Asia | 582 | 724 | 531 | 503 | 831 |
| Southeast Asia | 674 | 810 | 642 | 637 | 866 |
| Oceania | 56 | 492 | 167 | 436 | 41 |
| World | 266 | 876 | 521 | 662 | 156 |

**Source:** IFPRI calculation based on FAOSTAT 1999 and FAO 1997a.
**Notes:** (a) The total value of agricultural output was formed by weighting 134 primary crop and 23 primary livestock commodity quantities by their respective international agricultural prices for the 1989-91 period. (b) Agricultural land is the sum of cropland and permanent pasture.

Table 14
## Food Nutrition, Human Populations, and Agricultural Employment by Region, 1995–97 Average

| | Amounts Supplied by Agriculture[a] | | | | | | Population | | | Labor Force | | Total Population |
|---|---|---|---|---|---|---|---|---|---|---|---|---|
| | Calories | | Fat | | Protein | | Agri-cultural | Rural | Urban | Total | Agri-culture | |
| | (calories/ capita/ day: DES)[b] | (share of total DES)[b] | (grams/ capita/ day) | (share of total fat) | (grams/ capita/ day) | (share of total protein) | *(millions of people)* | | | | | |
| North America | 3,569 | (99.2) | 139.5 | (99.3) | 105.4 | (95.7) | 7.6 | 70.7 | 228.8 | 153.2 | 3.7 | 299.5 |
| Latin America and the Caribbean | 2,746 | (99.4) | 76.1 | (99.2) | 68.9 | (96.4) | 111.7 | 127.9 | 360.0 | 203.0 | 44.5 | 487.8 |
| Europe | 3,232 | (99.0) | 132.2 | (99.3) | 93.4 | (94.8) | 41.9 | 133.8 | 383.5 | 245.1 | 20.1 | 517.3 |
| Former Soviet Union | 2,776 | (99.2) | 73.3 | (99.2) | 79.9 | (95.8) | 49.7 | 92.9 | 198.9 | 146.8 | 23.5 | 291.8 |
| West Asia/ North Africa | 3,008 | (99.5) | 72.8 | (99.3) | 79.5 | (97.4) | 104.8 | 137.3 | 213.5 | 126.1 | 41.9 | 350.8 |
| Sub-Saharan Africa | 2,221 | (99.4) | 47.1 | (99.0) | 51.8 | (95.9) | 365.9 | 397.3 | 183.8 | 258.1 | 167.1 | 581.1 |
| East Asia | 2,783 | (98.0) | 68.1 | (97.8) | 69.6 | (90.2) | 871.6 | 895.6 | 539.8 | 839.2 | 517.8 | 1,435.4 |
| South Asia | 2,400 | (99.6) | 43.8 | (99.5) | 56.3 | (98.3) | 729.5 | 933.2 | 339.8 | 551.3 | 334.0 | 1,273.0 |
| Southeast Asia | 2,594 | (98.4) | 48.8 | (97.4) | 55.0 | (88.6) | 252.4 | 320.9 | 167.3 | 238.8 | 132.6 | 488.2 |
| Oceania | 2,940 | (99.0) | 112.5 | (99.1) | 89.1 | (94.7) | 5.3 | 8.6 | 20.3 | 13.6 | 2.5 | 28.9 |
| World | 2,732 | (99.0) | 70.3 | (98.6) | 69.3 | (94.1) | 2,540.4 | 3,118.2 | 2,635.6 | 2,775.1 | 1,287.8 | 5,753.7 |
| Percent of total population | | | | | | | 44.2 | 54.2 | 45.8 | 48.2 | 22.4 | 100.0 |

**Source:** Compiled from FAOSTAT 1999.
**Notes:** (a) Values exclude fish and other aquatic products. Values include livestock sources under both intensive and extensive (grasslands) grazing systems. (b) DES: Dietary Energy Supply.

## Figure 11

## Share of Major Food Groups in Total Dietary Energy Supply

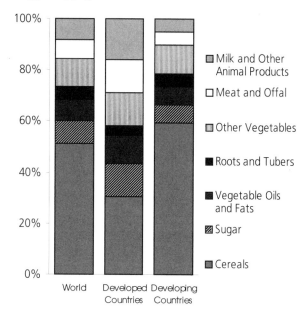

■ Milk and Other Animal Products

□ Meat and Offal

▥ Other Vegetables

■ Roots and Tubers

■ Vegetable Oils and Fats

▨ Sugar

■ Cereals

**Source:** FAO 1996:25.

but because of the higher proportion of confined animal production (or the limited amount of pasture) and a higher consumption of pig and poultry meat, Europe, South Asia, and Southeast Asia show markedly higher returns by this measure.

## NUTRITIONAL VALUE

The overriding purpose of agriculture is to provide an adequate and stable supply of food. Thus, a primary measure of the social value of agroecosystem outputs is their adequacy in satisfying human nutritional needs in terms of calories, proteins, fats, vitamins, and other micronutrients. The nutritional indicators adopted here are the per capita calories, fats, and proteins derived from agricultural sources. We use the per capita calorie indicator (also called the per capita Dietary Energy Supply, DES) as the primary indicator amongst these. As shown in Table 14, the global average DES supplied by agriculture is some 2,732 kilocalories per day (kcal/day), representing 99 percent of the total DES. In terms of protein, agriculture provides 69.3 grams per person, representing 94 percent of total protein intake (fish and other aquatic products make up most of the balance).

North America, Europe, and West Asia/North Africa enjoy the highest levels of DES, with over 3,000 kcal/day respectively. The lowest levels are found in Sub-Saharan Africa, South Asia, and Southeast Asia, with 2,200–2,600 kcal/day respectively. The regional disparities are even more marked in the case of proteins derived from agriculture, where North Americans con-

sume twice as much as Sub-Saharan Africans, 105 compared with 52 gram per day. Overall, the developed countries with 24 percent of the population consume 29 percent of the global DES, 34 percent of the global protein supply, and 43 percent of the global fat supply (FAO 1996:19). Actual per caput energy requirements vary by region because of systematic differences in body weights, metabolic rates, and human activity levels. Requirements for developing countries are typically around 2,100–2,200 kcal per day (FAO 1996:53).

Although, on the average, daily calorie requirements are exceeded by the DES in each region, per capita DES is broadly distributed about its mean value, highlighting that there are many people with inadequate food supply. In Sub-Saharan Africa, a staggering 33 percent of the population are undernourished, a significantly higher proportion than in Asia and Pacific, the next malnourished regions, although absolute numbers in Asia and Pacific are substantially higher (FAO 1999a:29). Although global food supply could provide adequate nutrition for the entire global population, a number of political and socioeconomic factors, including the limited capacity of the world's poor to purchase sufficient food, hamper progress to achieving these objectives.

Vegetable products contribute around 84 percent to overall DES (71 percent developed, 90 percent developing) (FAO 1996:23). This is mainly comprised of cereals, sugar, and vegetable oils and fats (see Figure 11). Trends reveal a significant increase in the importance of animal products, meat, milk, and eggs, in the developing world (see Box 2), as well as a general decline in the consumption of roots and tubers (with the notable exception of Sub-Saharan Africa where roots and tubers contribute 21 percent to DES (FAO 1996:25)).

## EMPLOYMENT AND INCOME

For rural households and communities—the majority of the world's population— agricultural labor represents the dominant, sometimes only, source of livelihood. In 1996, about 3.1 billion people (54 percent of global population) were living in rural areas, and of these about 2.5 billion were estimated to be living in agriculture dependent households (see Table 14). FAO estimates the agricultural labor force at 1.3 billion people (22 percent of the total population and 46.4 percent of the total labor force), but the proportions are highly variable. North America has only 2.4 percent of its labor force directly engaged in agriculture, while East, South, and Southeast Asia as well as Sub-Saharan Africa have between 55 and 65 percent.

In recognition of agriculture's importance as an income source for rural households, we have selected the annual value of agricultural production per agricultural worker as a proxy for the relative income potential of agricultural workers across the world's agroecosystems. The value of production per worker includes purchased inputs and other farm operating costs as

*Figure 12*

## Value of Production per Agricultural Worker, 1995-97 Average

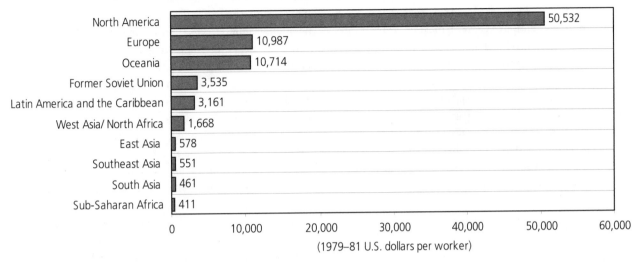

| Region | (1979–81 U.S. dollars per worker) |
|---|---|
| North America | 50,532 |
| Europe | 10,987 |
| Oceania | 10,714 |
| Former Soviet Union | 3,535 |
| Latin America and the Caribbean | 3,161 |
| West Asia/ North Africa | 1,668 |
| East Asia | 578 |
| Southeast Asia | 551 |
| South Asia | 461 |
| Sub-Saharan Africa | 411 |

**Source:** IFPRI calculation based on World Bank 2000 and FAOSTAT 1999.

well as labor. Globally, the value-added from agricultural production per member of the agricultural labor force is $1,027 per year (1989-91 U.S. dollars). The income measure is much higher in regions having both high overall productivity and low labor inputs *(see Figure 12)*. North America's income proxy of around $50,000 per worker per year is over 4 times greater than that of Europe. This figure contrasts starkly with the value of $411 (barely 1 dollar a day) per worker per year in Sub-Saharan Africa. The South Asia, Southeast Asia, and East Asia income proxies of $461, $551 and $578, respectively, are 12 to 40 percent higher but still low in absolute terms. The progression nearly triples with West Asia/North Africa posting $1,668 and Latin America and the Caribbean $3,161 per worker per year.

These figures need careful interpretation. Although absolute levels of value-added per worker are much higher in North America, much of the production value is derived from the greater use of capital and purchased inputs. Thus, the indicator value is not adjusted for the economic returns attributable to other production inputs, or for variations in the quality of labor. Another refinement would be to adjust values to reflect the purchasing power of a dollar across different regions. The actual incomes of most farmers in North America and Western Europe, in common with those in many countries around the world, are low in relation to other economic sectors, and are declining (USDA 2000). However, farmers in developing countries seldom receive income-enhancing transfers and subsidies commonly available to farmers in most industrialized countries.

## Enhancing the Capacity of Agroecosystems to Produce Food, Feed, and Fiber

Despite past successes, affordably feeding the current world population, and the more than 70 million people per year by which that population will continue to grow over the next 20 years, remains a formidable challenge. Area expansion options are limited, soil and water resources in agricultural areas are often already stressed, pesticide resistance is increasing, and growth in yields seems more difficult to achieve. Increasing agricultural productivity remains, therefore, a central agroecosystem development goal, but one in which the calculus of productivity is more broadly defined to encompass the use of physical and biological resource stocks and human health impacts.

The strategic cornerstones of enhanced productivity are likely to be the following: increasing genetic potential of crops and livestock; reducing biotic losses associated with pests, pathogens, and weeds; extending the range of abiotic adaptability (for example, by increasing salt and drought tolerance); and increasing the efficiency of natural resource use.

Both conventional breeding and biotechnology-based innovations will improve the yield potential of crops and livestock. Long-term productivity losses and environmental stresses caused by conventional intensification can also be eased by better crop selection and crop rotations, and by soil, fertilizer, pesticide, and water management practices that conserve soil and water, enhance soil fertility, and disrupt the development of pests, diseases, and pesticide resistance. IPM, an approach that combines biological and cultural control of pests with reduced and judicious use of pesticides, has proved extremely effective but is knowledge intensive. IPM is finding increasing favor with both commercial and smallholder farmers alike where specialist support is on hand to help adapt IPM practices to the constantly evolving field situation. However, because the use of pesticides will likely continue to grow in the foreseeable future, there is still a pressing need that they become increasingly target-specific and more rapidly and safely degraded. In developing countries, in particular, there is also a need to strengthen regulatory and enforcement mechanisms governing pesticide import, production and use, and to improve the training and protection of agricultural workers exposed to pesticides.

A common need underlying all of these strategies is that for improved knowledge. Ultimately this calls for continued investment in agricultural and natural resource research that can help design more productive and more environmentally beneficial farming systems and production technologies.

## Summary of Indicators and Data

At the global level, the indicators of food production levels and trends appear satisfactory. Compared to gloomy predictions of the world food situation by the end of the millennium, they could even be viewed as measures of considerable success. And despite slowing yield growth rates and increasing pesticide resistance, current expectations are that food production capacity can be kept in line with the growing demand in the global market (Pinstrup-Andersen et al. 1999).

There are, nevertheless, many underlying causes for concern. Perhaps the most telling are the enormous regional disparities that exist among the indicators of yield, nutrition, value, and income potential. These indicators are particularly troublesome since it is precisely in the areas of lowest productivity,

such as Sub-Saharan Africa, and of highest land degradation pressure such as much of Asia, Sub-Saharan Africa, and Central America, that the biggest advances in agricultural productivity need to be made. These areas have some of the highest population growth rates, and for the foreseeable future, they will continue to rely on national, and often local, sources to meet their food requirements.[13] Despite the growth and increasing liberalization of world trade, food imports today meet an average of only 10 percent of local food needs (McCalla 1999). Although trade expansion is likely, poverty is so widespread, foreign exchange so limited, and transport and marketing networks so poor in many of these countries, that purchasing imported food might not be an option for many.

At a regional and global scale of inquiry it is difficult to interpret available evidence in an integrated way, since there is so little information on actual land use patterns and practices. For example, little data exists on the application of organic nutrients and, when available, such data can be difficult to standardize among the different organic nutrient sources. Thus, although we have more (production, value, population) and less (technology, nutrients, pesticide, irrigation) reliable and complete sources of data, they are insufficient to determine how, say, nutrient, water, and pesticide inputs are likely to be combined by region, agroecosystem zone, and production system. Only by working at the appropriate scale and with spatial data that allow these pieces of information to be properly integrated—an explicit goal of the proposed Millennium Ecosystems Assessment—will it be feasible to better interpret the relationships between the capacity of agroecosystems to produce food versus their capacity to produce environmental goods and services.

# SOIL RESOURCE CONDITION

Soil, the principal medium for plant growth, is a thin layer of biologically active material lying over inert rock below, the result of complex processes of geologic weathering, nutrient cycling, and biomass growth and decay. Soil is the primary environmental stock that supports agriculture. Thus, good soil condition is central in determining the current state and future productive capacity of agroecosystems. It also greatly influences the provision of environmental services, including water flow and quality, biodiversity, and carbon. We discuss these roles in subsequent sections.

Since the emergence of agriculture, soil management has challenged farmers. Where land was abundant, farmers selected the best soils for production and used long fallow periods to restore soil physical and chemical properties lost by cultivation. With increasing population density, permanent agriculture evolved and farmers developed numerous techniques to sustain soil fertility, control the movement of soil and water, and improve soil characteristics favorable to crop cultivation. For millennia, farmers have used field rotation, leguminous and green manures, inorganic materials (such as lime and loess), and animal manures to recycle or concentrate plant nutrients in arable fields. The large-scale introduction of fertilizers, especially nitrogen and phosphorus, from natural or industrial sources has been a major source of yield growth. However, pro-

vision of mineral nutrients furnishes only partial restoration of the soil qualities essential to its sustainable use as a plant growth medium. Farmers and scientists have learned that soil productivity depends upon a range of interrelated factors including: soil organic matter, nutrient availability (particularly nitrogen, phosphorus, and potassium, but also micronutrients), water-holding capacity, soil reaction (pH), soil depth, salinity, the richness of the soil biota, and such physical characteristics as soil structure and texture.

This section first reviews the global pattern of inherent soil constraints, which highlights the spatial variability in soil resource endowment. Global evidence on the status of soil degradation in agricultural land is then assessed, followed by a discussion of more specific indicators of soil quality: soil organic matter, and soil nutrient balances. We conclude with a brief review of the prospects for enhancing agricultural soil's capacity to provide a range of goods and services.

## Global Patterns of Soil Constraints

We used the Fertility Capability Classification (FCC) approach (Sanchez et al. 1982; Smith 1989; Smith et al. 1997) to examine the inherent productive capacity of the world's soils. The approach interprets soil profile data to ascribe up to 20 "modi-

*Table 15*

## Share of PAGE Agricultural Extent[a] Affected by Major Soil Constraints[b]

| Region[c] | Share of PAGE Agricultural Extent | Free of Constraints | Poor Drainage | Low CEC | Aluminum Toxicity | Acidity | High P-Fixation | Vertisol | Low K-Reserves | Alkaline | Salinity | Natric | Shallow or Gravelly | Organic | Low Moisture Holding Capacity |
|---|---|---|---|---|---|---|---|---|---|---|---|---|---|---|---|
| North America | 12.2 | 29.3 | 17.6 | 0.0 | 18.1 | 28.2 | 0.0 | 1.4 | 12.8 | 6.8 | 0.2 | 1.9 | 4.0 | 1.5 | 8.9 |
| Latin America and the Caribbean | 17.3 | 12.4 | 11.3 | 7.2 | 30.9 | 17.0 | 13.1 | 3.8 | 39.3 | 5.4 | 3.2 | 4.8 | 10.1 | 0.3 | 12.9 |
| Europe | 9.2 | 18.2 | 18.3 | 1.2 | 11.7 | 51.1 | 0.0 | 1.6 | 0.0 | 8.6 | 1.0 | 1.3 | 8.0 | 2.2 | 14.7 |
| Former Soviet Union | 15.8 | 23.3 | 15.2 | 0.0 | 1.6 | 13.2 | 0.0 | 0.0 | 0.0 | 24.3 | 3.4 | 17.8 | 7.1 | 2.4 | 7.8 |
| West Asia/ North Africa | 3.1 | 20.4 | 3.2 | 0.1 | 2.4 | 14.2 | 0.0 | 4.9 | 1.2 | 30.1 | 4.5 | 0.6 | 20.9 | 0.0 | 1.6 |
| Sub-Saharan Africa | 16.1 | 7.0 | 9.5 | 15.9 | 20.5 | 28.6 | 4.3 | 6.2 | 29.7 | 2.7 | 1.3 | 2.3 | 11.4 | 0.5 | 23.2 |
| East Asia | 10.5 | 15.7 | 19.9 | 0.1 | 21.1 | 14.3 | 16.4 | 1.6 | 19.8 | 6.7 | 6.1 | 0.2 | 19.8 | 0.1 | 1.8 |
| South Asia | 8.4 | 13.4 | 6.8 | 0.7 | 5.0 | 32.2 | 0.2 | 19.5 | 9.0 | 9.6 | 8.3 | 0.5 | 9.4 | 0.2 | 7.9 |
| Southeast Asia | 5.6 | 6.2 | 21.7 | 2.3 | 37.9 | 32.6 | 8.1 | 3.1 | 43.5 | 0.6 | 0.9 | 0.3 | 6.4 | 3.3 | 6.0 |
| Oceania | 1.9 | 14.0 | 18.4 | 3.5 | 8.8 | 17.3 | 1.1 | 8.2 | 6.8 | 12.7 | 1.2 | 33.0 | 9.2 | 0.4 | 23.9 |
| *Total* | *100.0* | *16.2* | *14.0* | *4.2* | *17.2* | *24.6* | *5.2* | *4.3* | *18.6* | *9.5* | *3.0* | *5.1* | *10.0* | *1.1* | *11.3* |
| **Agroclimatic Zone[d]** | | | | | | | | | | | | | | | |
| Tropics/Arid and Semiarid | 14.4 | 8.4 | 7.9 | 11.8 | 7.2 | 29.6 | 1.2 | 16.5 | 11.9 | 4.1 | 2.6 | 3.9 | 13.3 | 0.2 | 20.8 |
| Tropics/Subhumid and Humid | 23.5 | 5.5 | 13.1 | 8.9 | 41.5 | 25.5 | 13.0 | 2.9 | 52.0 | 1.0 | 0.6 | 0.9 | 7.1 | 1.2 | 12.8 |
| Sub-tropics/ Arid and Semiarid | 9.4 | 24.1 | 5.6 | 3.2 | 1.1 | 13.6 | 0.0 | 4.3 | 1.3 | 25.3 | 11.8 | 7.6 | 15.6 | 0.0 | 13.9 |
| Sub-tropics/ Subhumid and Humid | 13.8 | 14.6 | 14.7 | 0.2 | 25.3 | 25.2 | 14.3 | 5.3 | 25.6 | 3.8 | 0.9 | 3.3 | 14.3 | 0.4 | 4.5 |
| Temperate/Arid and Semiarid | 20.1 | 25.5 | 13.1 | 0.1 | 1.1 | 9.6 | 0.0 | 0.1 | 0.1 | 23.9 | 5.5 | 14.9 | 9.8 | 1.4 | 5.0 |
| Temperate/ Subhumid and Humid | 18.0 | 23.1 | 24.3 | 0.6 | 14.3 | 39.5 | 0.3 | 0.5 | 5.7 | 6.7 | 0.9 | 1.3 | 5.1 | 1.9 | 13.4 |
| Boreal | 0.8 | 31.6 | 33.9 | 0.0 | 13.9 | 38.4 | 0.0 | 0.0 | 0.0 | 0.0 | 0.0 | 0.0 | 9.2 | 11.6 | 6.9 |
| *Total* | *100.0* | *16.2* | *14.0* | *4.2* | *17.2* | *24.6* | *5.2* | *4.3* | *18.6* | *9.5* | *3.0* | *5.1* | *10.0* | *1.1* | *11.3* |

**Source:** IFPRI calculation based on: (a) GLCCD 1998; USGS EDC 1999, (b) the fertility capability classification (FCC) applied to FAO's digital Soil Map of the World (FAO 1995; Smith 1989; and Smith et al. 1997), (c) country boundaries from ESRI 1996, and (d) FAO/IIASA 1999.

fiers," indicating soil constraints from an agricultural use perspective. We generated FCC modifiers for each of the 4,931 mapping units using data and analysis software included in FAO's Digital Soil Map of the World (FAO 1995). The proportional area of each soil constraint was then assessed for each five minute grid cell (approximately 10km$^2$ at the equator) of the soil map.[15]

Map 10 indicates which of the FCC constraints, including no constraint, is dominant (occupies the greatest proportional area) at each location within the PAGE global agricultural extent. Table 15 summarizes the area within the PAGE agricultural extent affected by soil constraints.

Acidity, defined in the FCC as a soil pH between 5.0 and 6.0 (Sanchez et al. 1982 quoted in FAO 1995), is the most common

global soil constraint, affecting a quarter of the PAGE agricultural extent. A soil pH of around 6.0-7.0 increases the availability of nutrients and promotes beneficial microbial activity. Acid soils tend to be saturated with exchangeable aluminum. In some 17 percent of the agricultural extent, aluminum saturation is so high as to be toxic to plants. These problems are particularly acute in the highly weathered soils of the humid tropics, but can also be induced by long-term use of ammonia-based fertilizers. Other constraints linked to acid soils include a high capacity to "fix" natural or applied phosphorus, making phosphorus unavailable to plants, as well as low reserves of other nutrients, such as potassium. Phosphorus fixation is prevalent in only 5 percent of agricultural lands, but is a serious concern in parts of Latin America, East Asia, Southeast Asia, and Sub-

*Table 16*
## PAGE Agricultural Extent[a] by Major Climate,[b] Slope,[c] and Soil Constraints[d]

| Major Climate[b] | 0 to 8 Percent Slope[c] | | | 8 to 30 Percent Slope[c] | | | Greater than 30 Percent Slope[c] | | |
| | Occurrence of Soil Constraints[d] | | | | | | | | |
| | Predominant | Moderate | Low | Predominant | Moderate | Low | Predominant | Moderate | Low |
| | (>= 70%) | (30-70%) | (<30%) | (>= 70%) | (30-70%) | (<30%) | (>= 70%) | (30-70%) | (<30%) |
|---|---|---|---|---|---|---|---|---|---|
| | *(percentage)* | | | | | | | | |
| **Tropics** | | | | | | | | | |
| Arid/Semiarid | 7.8 | 0.5 | 0.1 | 4.9 | 0.5 | 0.1 | 0.4 | 0.2 | 0.0 |
| Subhumid/Humid | 13.0 | 0.3 | 0.1 | 7.8 | 0.3 | 0.1 | 1.5 | 0.2 | 0.1 |
| **Subtropics** | | | | | | | | | |
| Arid/Semiarid | 3.9 | 0.7 | 0.9 | 2.2 | 0.4 | 0.2 | 0.8 | 0.3 | 0.0 |
| Subhumid/Humid | 4.0 | 0.6 | 0.7 | 4.7 | 0.5 | 0.3 | 2.7 | 0.2 | 0.1 |
| **Temperate** | | | | | | | | | |
| Arid/Semiarid | 8.7 | 0.9 | 2.2 | 4.3 | 1.4 | 1.2 | 0.8 | 0.6 | 0.1 |
| Subhumid/Humid | 7.2 | 1.2 | 1.9 | 5.6 | 0.8 | 0.5 | 0.7 | 0.2 | 0.0 |
| Boreal | 0.2 | 0.1 | 0.0 | 0.1 | 0.4 | 0.0 | 0.0 | 0.0 | 0.0 |
| Total | 44.7 | 4.2 | 6.1 | 29.6 | 4.2 | 2.3 | 6.9 | 1.7 | 0.3 |

**Source:** IFPRI calculation based on: (a) GLCCD 1998; USGS EDC 1999, (b) FAO/IIASA 1999, (c) FAO/IIASA 1999 based on USGS 1998, and (d) the fertility capability classification (FCC) applied to FAO's Digital Soil Map of the World (FAO 1995; Smith 1989; and Smith et al. 1997).

Saharan Africa. Low potassium reserves are often found in the same regions, but are more extensive, occupying some 19 percent of the agricultural extent.

Poorly drained (hydromorphic) soils are found in about 14 percent of the agricultural extent largely in areas where physiography promotes flooding, high groundwater tables, or stagnant surface water. While problematic for many agricultural purposes, they are suited to rice cultivation and seasonal grazing (Bot and Nachtergaele 1999).

Although agricultural drylands are known to be underrepresented in the PAGE agricultural extent, constraints often related to drier environments feature significantly in Table 15. Saline soils—those with excess soluble salts—are found on around 3 percent of the agricultural extent, and natric (sodium rich) soils on around 5 percent of these lands, the latter particularly prevalent in Oceania and the Former Soviet Union. Although salinity presents problems of toxicity to most crops, sodicity inhibits infiltration and root development. Both salinity and sodicity are associated with dryer areas and more alkaline (basic) soils. Excessively alkaline soils, defined in the FCC schema as having a pH of greater than 7.3 (Sanchez et al. 1982 quoted in FAO 1995) occupy around 10 percent of the agricultural extent and, as with soil acidity, inhibit the availability of plant nutrients.

Focusing only on the proportion of PAGE agricultural extent free from soil constraints *(see Map 11)* provides another per-

spective on the spatial distribution of soil quality. Only 16 percent of agricultural soils are free from constraints. About 60 percent of those favored soils lie in temperate areas, particularly the midwestern United States, central western Canada, Russia, central Argentina, Uruguay, southern Brazil, northern India, and northeast China, while only 15 percent lie within the tropics.

To link soil constraints to broader climatic and physiographic factors, we defined three classes of constraints: predominantly constrained (over 70 percent of land having soil constraints), moderately constrained (30-70 percent with constraints), or relatively unconstrained (under 30 percent with constraints). We combined these extents with major climate, moisture availability and slope maps to assess the relative disposition of favored (flat, well-watered, fertile) compared to less-favored (sloping, drier, less fertile) lands within the PAGE agricultural extent. Table 16 summarizes this distribution.

Farmers generally prefer flatter lands without important soil quality constraints. Globally, in the semiarid, subhumid, and humid subtropics and the subhumid temperate zones, over three quarters of all available flat lands (defined as those with under 8 percent slope) without soil constraints (under 30 percent constrained) were found within the agricultural extent. However, large areas of steeper and soil-constrained land are also under cultivation. In some cases, these soils have positive attributes that make them preferred (for example, steeper slopes provide better drainage on heavier soils; vertisols are difficult to work manually but are quite fertile). For historical and political reasons, some ethnic groups are concentrated in regions with high soil constraints. In many cases, population growth and land scarcity in higher-quality agricultural regions have resulted in large groups of farmers migrating to regions having lower quality soils.

## Soil Quality Status and Change

Natural weathering processes and human use have brought about continuous changes in soil quality. The deterioration over time of key soil attributes required for plant growth, or for providing environmental services, constitutes "degradation." The principal soil degradation processes are erosion by water or wind, salinization and waterlogging, compaction and hard setting, acidification, loss of soil organic matter, soil nutrient depletion, biological degradation, and soil pollution.

Climate plays a critical role in the degradation process. As temperatures increase, decay of organic matter is greatly accelerated, particularly in frequently tilled soils. Increased temperatures coupled with decreased precipitation tend to intensify degradation, making many soil-conservation practices less effective. Additionally, the potential for wind and water erosion is generally greater in warmer areas. Erosion is further increased in more arid areas because of lower inherent organic-matter levels and less natural vegetation. Retaining crop residues, a key feature in sustainable soil management, is much more difficult in hot and arid areas. In tropical areas, continuous cropping often results in quite rapid degradation and consequent productivity decline through increased chemical acidity, loss of essential plant nutrients, and structural collapse of the soil. Subsequently, the soil's capacity to form stable aggregates is reduced because soil organic matter, the binding material, has been lost (Stewart, Lal, and El-Swaify 1991).

Different land management practices are associated with different types and degrees of soil degradation. For example, salinization is linked with intensification on irrigated land and clearing of deep-rooted vegetation, compaction with mechanized farming in high-quality rainfed lands, nutrient depletion with intensification of marginal lands, water erosion with clearing and extensive management of marginal rainfed lands, and soil pollution with periurban agriculture (Scherr 1999b).

The diversity of causes, processes, and consequences of soil degradation presents challenges in defining useful indicators to describe its status. This section reports briefly on the only global attempt to define such indicators, the Global Assessment of Soil Degradation (GLASOD, Oldeman et al. 1991a) and interprets its findings within the PAGE agricultural extent.

For GLASOD, regional experts used a standardized assessment framework and regional (1:10M scale) maps to judge the prevalence of four types of human-induced degradation: water erosion, wind erosion, physical degradation, and chemical degradation. For each degradation type, the GLASOD experts assessed both the proportional area affected by degradation (*extent*) and the scale of degradation on affected areas (*degree*). In terms of degradation extent, the GLASOD results suggested that about 23 percent of all used land was degraded to some degree: 38 percent of cropland; 21 percent of permanent pasture; and 18 percent of forests. Of the degraded lands, about 38 percent was lightly degraded; 46 percent moderately degraded (suggesting significantly reduced agricultural productivity and partial destruction of original biotic functions); and 16 percent was strongly to extremely degraded, no longer suitable for agricultural use. The GLASOD analysis attributed around 35 percent of the extent of human-induced degradation to overgrazing and about 28 percent to other agricultural-related land management. About 29 percent was also attributed to deforestation (Oldeman 1994:111-116 and Oldeman et al. 1991b) and it is likely that a significant share of the cleared land was used for agricultural purposes.

For mapping purposes GLASOD degradation extent and degree attributes were combined to form degradation *severity* classes (Oldeman et al. 1991a:15). We overlaid the GLASOD map of degradation severity with the PAGE agricultural extent map to obtain a spatial perspective of the correspondence of degradation and agriculture *(see Map 12)*. GLASOD mapping units falling within the PAGE agricultural extent tend to have higher severity classes than are found across all land use types (agriculture, grazing and forest land combined). The overlay reveals that 35 percent of the PAGE agricultural extent coincides with GLASOD mapping units classed as not degraded or as having low severity degradation, while over 40 percent of the agricultural extent coincides with mapping units whose degradation severity is high or very high.[15] This interpretation, however, gives an overly pessimistic view of soil degradation severity in agricultural lands. Both degradation and agriculture are defined as occupying only a proportion of any given area. In reality, the amount to which the degraded areas within GLASOD mapping units physically overlap agricultural areas within the PAGE agricultural extent, is unknown. For example, a degradation severity class of "very high" is assigned to an entire GLASOD mapping unit if as little as 10 percent of its *extent* has an extreme *degree* of erosion *(see legend of Map 12)*. Since the PAGE agricultural extent includes areas that are up to 70 percent non-agricultural *(see Map 1)*, it is quite conceivable that the 10 percent of the area degraded is non-agricultural and, consequently, that the agricultural area has no—rather than very high—degradation.

To avoid this bias, we characterized GLASOD degradation mapping units not by severity class but by their component extent and degree attributes. We then tabulated PAGE agricultural extent areas that coincided with different combinations of

*Table 17*

## Distribution of PAGE Agricultural Extent by GLASOD Mapping Unit

| Degradation Degree by GLASOD mapping unit | Degradation Extent (degraded share of GLASOD mapping unit) | | | | | | |
|---|---|---|---|---|---|---|---|
| | 0 | 1-5 | 6-10 | 11-25 | 26-50 | >50 | All |
| | *(percent of agricultural extent)* | | | | | | |
| None | 14.6 | | | | | | 14.6 |
| Light | | 12.2 | 8.2 | 10.1 | 2.6 | 0.3 | 33.5 |
| Moderate | | 6.6 | 11.6 | 14.1 | 8.9 | 1.7 | 42.9 |
| Strong | | 1.2 | 1.1 | 2.9 | 2.0 | 1.3 | 8.4 |
| Extreme | | 0.2 | 0.0 | 0.3 | 0.0 | 0.0 | 0.5 |
| All | 14.6 | 20.2 | 20.9 | 27.4 | 13.4 | 3.4 | 100.0 |

**Source:** IFPRI calculation based on GLASOD (Oldeman et al. 1991a) and the PAGE agricultural extent (GLCCD 1998; USGS EDC 1999).
**Notes:** Relationships between the degradation of soil and its capacity to provide goods and services are complex. Depending on soil type, for example, a given depth of soil erosion could have negligible to very significant consequences for crop productivity.

degradation extent and degree from the GLASOD map. The results of this tabulation are shown in Table 17. In terms of degradation extent, around 35 percent of PAGE agricultural extent has 5 percent or less of its area degraded. About half of the PAGE agricultural extent corresponds with GLASOD mapping units that have 6 to 25 percent of their extent degraded. Less than 5 percent of the PAGE agricultural extent corresponds with degradation mapping units where more than half the area is degraded. In terms of degradation degree, about 48 percent of the agricultural extent is only lightly degraded or not degraded, while 9 percent is strongly or extremely degraded.

The consequences of different types and degrees of soil degradation on agricultural productivity and the provision of environmental services vary widely and are not always well understood. However, these global soil degradation estimates for agricultural lands are cause for significant concern. The picture they paint calls, at the very least, for a greater sense of urgency with regard to more reliable monitoring of the location, extent, degree, and impact of soil degradation. Such information is an essential prerequisite to assessing the priority and scale of appropriate remedial measures.

### SOIL DEGRADATION IN SOUTH AND SOUTHEAST ASIA

The most comprehensive regional data on soil degradation is from the Assessment of Human-Induced Soil Degradation in South and Southeast Asia (ASSOD). The ASSOD study used a methodology similar to GLASOD, but was nationally representative (1:5M scale) and checked against available national data. The definition of "degradation" is different from GLASOD, as it represents expert assessment of the degree of yield loss associated with soil quality change. Thus, significant deterioration in key soil qualities that did not result in yield change was assessed as "no or negligible degradation" (van Lynden and Oldeman 1997).

ASSOD found that agricultural activity had led to degradation on 27 percent of all land, and deforestation on 11 percent since around the middle of the 20th century; overgrazing played a minor role. The more detailed ASSOD study showed more degraded land than the GLASOD study showed but, as in the case of erosion by water, often to a lesser degree. ASSOD did, however, show significantly greater extents of light to moderate loss of soil fertility, and strong and extreme terrain deformation by water and wind than did GLASOD *(see Table 18)*.

ASSOD collaborators also provided data on types of farm management for nearly half of the degraded land. They found little association between land management and degradation: 38 percent of degraded lands were under a high level of management, 36 percent under medium management, and 25 percent under low management (defined as "traditional" systems existing for more than 25 years). In recent years, however, degradation had increased more often under low and medium management (van Lynden and Oldeman 1997:26-27).

Map 13 presents the ASSOD data on severity of soil degradation, within the PAGE agricultural extent. It highlights the geographic concentration of areas with the most serious degradation. The underlying data show the following: chemical deterioration and salinization are found in the same areas; fertility decline and water erosion are more widespread in China; and water erosion is widespread, especially in the agricultural areas of Thailand, India, and China.

## Soil Organic Matter

The presence of soil organic matter (SOM), those parts of the soil that originated from plants and animals, is one of the single most important measures of soil quality and, hence, of agroecosystem condition. The beneficial attributes that SOM imparts include the following:

*Table 18*

## Degraded Lands[a] within the PAGE Agricultural Extent[b] for South and Southeast Asia

| Dominant Degradation Type | Impact | | | |
|---|---|---|---|---|
| | None/Low | Moderate | Strong | Extreme |
| | *(percent of degraded PAGE agricultural land)* | | | |
| Water Erosion[c] | 16.2 | 17.4 | 20.1 | 9.1 |
| Wind Erosion[d] | 0.1 | 1.6 | 1.3 | 2.6 |
| Chemical Degradation[e] | 9.2 | 2.8 | 3.1 | 7.3 |
| Physical Degradation[f] | 0.4 | 3.0 | 5.4 | 0.4 |
| Total | 25.9 | 24.8 | 29.8 | 19.5 |

**Source:** IFPRI calculation based on: (a) the Assessment of the Status of Human-Induced Soil Degradation in South and Southeast Asia (ASSOD: van Lynden and Oldeman 1997), and (b) GLCCD 1998.
**Notes:** (c) Water erosion includes: loss of topsoil and terrain deformation; (d) wind erosion includes: loss of topsoil by wind action, terrain deformation, and overblowing; (e) chemical degradation includes: fertility decline and reduced organic matter content, salinization/alkalinization, dystrification/acidification, eutrophication, and pollution; (f) physical degradation includes: compaction, crusting and sealing, waterlogging, lowering of the soil's surface, loss of productive function, and aridification.

♦ stabilizes and holds together soil particles—thus reducing erosion;

♦ provides a source of carbon and energy for soil microbes;

♦ improves the soil's ability to store and transmit air and water;

♦ stores and supplies nutrients such as nitrogen, phosphorus, and sulphur;

♦ maintains soil in an uncompacted condition and makes the soil easier to work;

♦ retains carbon from the atmosphere and other sources;

♦ retains nutrients (e.g., calcium, magnesium, potassium) by providing ion exchange capacities; and,

♦ serves to reduce the negative environmental effects of pesticides and other pollutants.

Through such effects, soils rich in organic matter not only yield more food, because they are more productive, but also enhance soil biodiversity, store carbon, regulate surface water flows, and improve water quality—all key goods and services of relevance to this study. Factors influencing SOM formation in agroecosystems include crop type and rotation practices, crop residue availability, application of organic and inorganic nutrients, and the population and vigor of soil biota. Other conditions being equal, organic matter accumulates most at low temperatures in more acidic soils and under anaerobic conditions (Batjes 1999).

Land conversion into agriculture is a primary cause of SOM decline. SOM levels tend to fall as litter formation rates are decreased, organic matter is oxidized from increased tillage, and soil erosion increases. Good farming can slow the decline and can establish SOM formation and decomposition equilibrium rates that maintain long-term soil quality and productivity. This is, however, much more easily achieved in temperate

rather than in tropical regions, where higher temperatures, and sometimes higher erosion and leaching rates, accelerate SOM loss. Tiessen et al. (1994:784) found that 65 years of cultivation of a Canadian prairie reduced the soil carbon content by almost 50 percent, while only 6 years of cultivation in a Brazilian semi-arid thorn forest reduced the soil carbon content by 40 percent, and just three years of slash-and-burn cultivation of a Venezuelan rain forest soil reduced litter layer carbon by around 80 percent and the soil carbon by around 30 percent. Ironically, SOM plays a critical role in the quality of highly-weathered tropical soils where the lack of inorganic nutrients, and the limited natural regeneration, place greater reliance on the decomposition of litter or of applied organic matter to enhance soil fertility. Simply applying more inorganic fertilizer in such situations is seldom effective as a minimum threshold amount of SOM is required to bind incoming inorganic nutrients.

The indicator of SOM used here is the organic carbon content of soil[16] to a depth of 100 centimeters. The data set used for this indicator, described in the Carbon Services section, combines carbon data from soil profiles (Batjes 1996; Batjes 2000) with maps of soil distribution (FAO 1995), in order to estimate the organic carbon storage of the world's soils. Table 19 summarizes the distribution of soil organic carbon storage by region and Table 26 shows the distribution by agroecological zone.[17] The data clearly demonstrates that soil carbon densities are significantly higher in temperate latitudes (except for those tropical areas, mainly in Southeast Asia, that contain peat soils).

Although the soil organic carbon indicator is usable at multiple scales, routinely collected in soil surveys, and has standardized laboratory measuring techniques, there remain some practical questions over its use. One is the selection of soil depth over which the indicator is assessed. From a carbon sequestration and storage perspective (for which the indicator is used in

*Table 19*

**Soil Organic Carbon Storage (0–100 centimeters soil depth) within the PAGE Agricultural Extent[a]**

| Region[c] | PAGE Agricultural Extent[a] | | Total SOC[b] | SOC Density |
|---|---|---|---|---|
| | *(m sq km)* | *(% total)* | *(GtC)* | *(MtC/ha)* |
| North America | 4.4 | *12.2* | 54 | 122 |
| Latin America and the Caribbean | 6.2 | *17.3* | 59 | 95 |
| Europe | 3.3 | *9.2* | 49 | 146 |
| Former Soviet Union | 5.7 | *15.8* | 66 | 116 |
| West Asia/North Africa | 1.1 | *3.1* | 8 | 71 |
| Sub-Saharan Africa | 5.8 | *16.1* | 42 | 77 |
| East Asia | 3.8 | *10.5* | 32 | 85 |
| South Asia | 3.0 | *8.4* | 25 | 83 |
| Southeast Asia | 2.0 | *5.6* | 26 | 126 |
| Oceania | 0.7 | *1.9* | 5 | 78 |
| Total PAGE Agricultural Extent | 36.2 | | 369 | 102 |
| Total Land Area[d] | 130.4 | | 1,555 | |
| Agricultural Percentage of Total | *27.8* | | *23.7* | |

**Source:** IFPRI calculation based on: (a) GLCCD 1998; USGS EDC 1999, (b) FAO 1995, Batjes 1996 and Batjes 2000. *(see note)*, and (c) ESRI 1996. (d) FAOSTAT 1999.
**Note:** Batjes (1996) estimated the average soil organic carbon (SOC) content at a depth of 100cm by soil type based on over 4,000 individual soil profiles contained in the World Inventory of Soil Emission Potentials (WISE) database compiled by the International Soil Reference and Information Centre (Batjes and Bridges 1994). The authors calculated the global estimate of SOC storage by applying Batjes' (1996 and 2000) SOC content values by soil type to the soil type area share of each 5 x 5 minute unit of the Soil Map of the World (FAO 1995). SOC data for Greenland and Antarctica were largely incomplete and were excluded.

the section on Carbon Services), the goal is to assess total carbon within the entire soil profile because carbon found below the plow layer is more permanently stored. For agricultural productivity purposes, however, it is SOM dynamics in litter and the upper soil horizons (to 30cm) that are likely more critical. Furthermore, the underlying soil profile data is from various (often distant) points in time and may be unrepresentative of past and current land uses. The long-term monitoring of SOM is increasingly viewed as an important means of measuring progress toward achieving sustainable agriculture and of keeping abreast of degradation trends.

## Soil Nutrient Dynamics

As production intensifies, so does the challenge of maintaining balanced soil nutrient conditions. Shorter fallow periods, greater planting densities, and higher grain yields place greater demands on soil nutrients. Net nutrient depletion ("mining") will occur if extraction rates consistently exceed replacement and regeneration rates. Conversely, if nutrients are applied in excess of requirements, residues often pass into surface- and groundwater resources *(see Box 4 in the section on Water Services)*. Nutrient mining directly affects soil productivity and, sooner or later, will require a change in management practices, such as fallow, increased nutrient inputs, less demanding cropping systems, or abandonment. Leaching of nutrient residuals

is a production externality whose impacts are most apparent to downstream water users and to those concerned with the welfare of downstream aquatic ecosystems. The sign, size, and trend of annual nutrient balances can, thus, serve as an indicator of the changing productive capacity of agroecosystems (as well as for their nutrient leaching potential).

On behalf of the PAGE study, the International Fertilizer Development Center (IFDC) made a country- and crop-specific assessment of nutrient balances for Latin America and the Caribbean (Henao 1999). The analysis brought together data on inorganic fertilizer consumption, crop-specific fertilizer-application, organic-fertilizer use, and crop-residue recycling, as well as on nutrient extraction rates. Average annual nutrient balances were computed for two time periods, 1983-85 and 1993-95, as the difference between the sum of the nutrient inputs and outputs. Nutrient inputs comprised the following:

♦ Mineral fertilizer applied in kilograms of nitrogen (N), phosphorus ($P_2O_5$), and potassium ($K_2O$) per hectare (together referred to as NPK);

♦ Organic fertilizer applied as manure or animal residue expressed as kilograms of NPK per hectare;

♦ NPK derived from crop residues left on the soil after harvest and estimated as kilograms of NPK per hectare; and

♦ Nitrogen fixation by soybeans and pulse crops, expressed in kilograms of NPK per hectare.

*Table 20*

## Nutrient Balances by Crop: Latin America and the Caribbean

| Region[a] | Crop | | | | | | | | | |
|---|---|---|---|---|---|---|---|---|---|---|
| | Wheat | Rice | Maize | Sorghum | Potato | Cassava | Beans | Soybean | Other | All |
| | *(kg NPK[b] per hectare per year)* | | | | | | | | | |
| **1983-85 Average** | | | | | | | | | | |
| Mesoamerica | -198 | -89 | -17 | -120 | 15 | -54 | 14 | -88 | -32 | -39 |
| Caribbean | | -197 | -70 | -120 | -67 | -23 | -11 | | -39 | -67 |
| Andean | -77 | -110 | -47 | -32 | -60 | -63 | -36 | -114 | -51 | -57 |
| Southern Cone | -101 | -46 | -89 | -162 | -77 | -48 | 4 | -27 | -68 | -65 |
| *LAC Average* | *-111* | *-62* | *-61* | *-133* | *-60* | *-49* | *5* | *-30* | *-58* | *-59* |
| | | | | | | | | | | |
| **1993-95 Average** | | | | | | | | | | |
| Mesoamerica | -199 | -105 | -49 | -111 | -112 | -126 | 4 | -86 | -11 | -43 |
| Caribbean | | -170 | -33 | -85 | 12 | -20 | 35 | | -10 | -41 |
| Andean | -79 | -73 | -37 | -8 | -6 | -57 | -47 | -165 | -28 | -40 |
| Southern Cone | -83 | -72 | -115 | -161 | -21 | -31 | 7 | -24 | -50 | -59 |
| *LAC Average* | *-96* | *-77* | *-86* | *-108* | *-18* | *-35* | *4* | *-28* | *-37* | *-54* |

**Source:** IFPRI calculation based on Henao 1999.
**Notes:** (a) Mesoamerica includes Costa Rica, El Salvador, Guatemala, Honduras, Mexico, Nicaragua, and Panama. The Caribbean includes Cuba, Dominican Republic, Haiti, and others. Andean countries include Bolivia, Colombia, Ecuador, Peru, and Venezuela. The Southern Cone includes Argentina, Brazil, Chile, Paraguay, and Uruguay. (b) NPK: Nitrogen (N), phosphorus ($P_2O_5$), and potassium ($K_2O$).

Nutrient outputs consisted of the following:

♦ Nutrient uptake in grain or main crop product in kilograms of NPK per hectare; and

♦ Nutrient uptake in main crop residue in kilograms of NPK per hectare. Depending on the crop and country, some proportion of this extraction was assigned for recycling *(see nutrient inputs)*.

The results of this analysis, summarized by subregion in Table 20, suggest that for most crops and cropping systems in Latin America and the Caribbean (LAC) the nutrient balance is significantly negative, although depletion rates appear, in general, to be declining. Across all crops, the nutrient stocks of the region's soils were being depleted by around 54 kg NPK ha$^{-1}$yr$^{-1}$ in 1993-95. This rate was down 8 percent from 59 kg NPK ha$^{-1}$yr$^{-1}$ in 1983-85. Stoorvogel and Smaling (1990) reported net losses of around 49 kg NPK ha$^{-1}$yr$^{-1}$ for Sub-Saharan Africa where yields and nutrient application rates tend to be lower. Their estimates included such factors as erosion. Comparing LAC nutrient balances between 1984 and 1994 indicates that nutrient depletion under rice and maize is accelerating. On a subregional basis, the Andean, Caribbean, and Mesoamerica regions[18] had average depletion rates in 1994 of 40-43 kg ha$^{-1}$yr$^{-1}$ and the Southern Cone countries 59 kg ha$^{-1}$yr$^{-1}$. These rates are 30, 39, 10, and 9 percent lower, respectively, compared to 1984.

In order to develop nutrient balance maps, we applied the nutrient balances for each of the major Latin America and the Caribbean cereals: wheat, rice, maize, and sorghum (which to-gether account for around 36 percent of the arable crop area) onto maps of crop distribution for each cereal estimated by IFPRI (Sebastian and Wood 2000). The four resulting nutrient balance maps were aggregated to obtain a soil nutrient balance map for lands under cereals *(see Map 14a)*. The predominance of negative balances for the period 1993-95, particularly in Argentina, is clearly visible. Positive balances appear for Venezuela and Ecuador because of the positive nutrient balance assigned to maize in those countries. The major areas of significantly negative nutrient balance in cereal production are found in Buenos Aires province in Argentina, as well as in other parts of Argentina and in the Brazilian cerrados.

### HOT SPOTS AND BRIGHT SPOTS IN LATIN AMERICA AND THE CARIBBEAN PRODUCTIVITY TRENDS

Information on cereal yield trends (1975-95) was combined with that of nutrient balances *(see Map 14b)*. Long-term cereal yield trends are positive in Argentina, Chile, much of southern Brazil, Uruguay, Venezuela, and much of Mexico. Cereal yields were stationary or decreasing in northeast Brazil, Paraguay, most parts of the Andean countries, and many Caribbean countries.

The superposition of yield trend and nutrient balance maps provides insights into the spatial pattern of agricultural productivity trends *(see Map 14c)*. Potential bright spots were defined as stable or increasing yields with positive or only marginally negative nutrient balances (0 to –25 kg/ha per year), noting that Latin America and the Caribbean fertilizer application rates are seldom sufficient to pose major water pollution threats. Potential hot spots were identified as areas in which yields are decreasing and the nutrient deficits are greater than 25 kg/ha per year, or where yields are stable but the nutrient deficit is greater than 100 kg/ha per year.

The mapped index draws attention to ongoing deterioration of the biophysical production capacity of agroecosystems, a situation that is not sustainable in the long term. At some time (unspecified here as we have insufficient data to assess the total nutrient stock of soils), land must either be abandoned, left as fallow to naturally restore its fertility, or additional investments must be incurred to replenish nutrient stocks.

## Interpreting Regional and Global Soil Quality

For the foreseeable future, human food supply will continue to depend upon maintaining the productive potential of soil. Water and carbon cycles will depend on maintaining the ecosystem services of soils. It is difficult to relate the impacts of soil erosion and changing soil quality to agricultural supply, ecosystem condition and human welfare.

Rough estimates of agricultural productivity loss, based on GLASOD data, suggest cumulative productivity loss from soil degradation over the past 50 years to be about 13 percent for cropland and 4 percent for pastureland (Oldeman 1998:4). Crop loss yields in Africa from 1970-90 resulting from water erosion alone are estimated to be 8 percent (Lal 1995). The economic losses from soil degradation range from under 1 to 7 percent of agricultural gross domestic product (AGDP) in South and Southeast Asia (Young 1994), and under 1 to 9 percent of AGDP in eight African countries (Bøjö 1996). Unfortunately, most of these estimates are based on crude models using the areal extent of degradation, average aggregate rates of degradation extrapolated from experimental plots, and average values of lost yield. They fail to highlight the variations between regions or land use systems in soil conditions that are crucial to guiding policy and action.

Analyses at the subregional level are much more reliable, as they are able to take into account variations in soil, soil management, and farm economic conditions (for examples, see Ali and Byerlee forthcoming; Huang et al. 1996; Lindert forthcoming). Many farming communities have been successful in modifying their resource management practices in response to levels of degradation that might threaten their livelihoods or in response to new opportunities to capitalize the value of land improvements (Boserup 1965; Scherr 1999b). The indicators currently available, however, do not allow us to monitor such trends. Furthermore, data that would permit analysis of the capacity of different soils to recover from degradation, and hence the long-term risk to food security, is largely unavailable for many soil types.

## Enhancing the Capacity of Soils to Provide Goods and Services

In considering the significance of soil constraints, it is important to recognize that many plants are adapted to tolerate some constraints if they are not too severe, even though higher yields are likely to be obtained in unconstrained soils. Irrigated paddy rice production systems create a unique flooded soil environment for the rice plant that overcomes many constraints inherent in the original soils. Tree and shrub species, such as *Atriplex,* grow in saline soils, *Eucalyptus* in poorly drained soils, *Tithonia* in phosphorus-deficient soils, tea in acid soils, and olives in dry soils. Crop-improvement programs around the world are actively seeking to develop germplasm that can produce acceptable yields, even with soil constraints such as moderate acidity and salinity. In many cases, farmers can grow crops in less hospitable environments by amending, draining, or specially preparing the soil. If land with better soils is unavailable, they may simply tolerate relatively low yields to obtain essential crop products.

Raising the productive capacity of soils, those inherently constrained as well as those whose capacity is being reduced by degradation, is an increasingly strategic area of agricultural research. On the one hand are the goals of arresting soil degradation; on the other is the growing belief that increased understanding of nutrient recycling, soil organic matter dynamics, and soil biology will help design of more efficient production systems that accentuate synergies between abiotic and biotic resources. Early efforts, that increasingly try to link scientific and local innovations, include greater use of nitrogen-fixing legume crops or trees; incorporation of crop residues, and animal and green manure, and soil treatments with locally available materials, such as leaf litter, lime, and rock phosphates. Researchers are developing lower-cost conservation practices, such as contour vegetative strips producing cash crops, that increase short-term productivity and income and, thus, encourage farmer adoption. Many "bright spots" where soil quality is improving have been identified (Scherr and Yadav 1995), and, in the past decade, public and NGO investment in agricultural land protection, rehabilitation, and improvement has increased significantly. Major multilateral efforts to develop and apply such integrated soil, water, and nutrient management approaches include the World Bank-coordinated Soil Fertility Initiative for Africa (SFI 2000) and the CGIAR's Soil, Water, and Nutrient Management Program (SWNM 2000). The Convention to Combat Desertification (UNCCD 2000) promotes dryland rehabilitation. There may be future avenues to mobilize private investment in land improvement for carbon emission offsets through the Kyoto Protocol of the Climate Change Convention (UN FCCC 2000). Ultimately, whether farmers invest to maintain and improve soil quality will depend on the economic prospects of agriculture, and the mobilization of investment resources.

Although such initiatives face significant challenges, the potential gains from improving soil quality are more far-reaching than the increased welfare of food producers and consumers. Improved soil quality will also enhance agroecosystem capacity to maintain water flow regulation, water quality, crop and soil biodiversity, and carbon storage services.

## Summary of Indicators and Data

The analysis of inherent soil constraints presented a relatively static picture of soil resource quality domains within which various forms of degradation are occurring. Therefore, they are not indicators of soil capacity in and of themselves, but are useful to stratify observations of selected indicators such as SOM and

nutrient balance, across broader, similar areas (using soil process or pedotransfer models). The SOM measure shown here was derived from the same type of soil survey data as the soil constraints analysis and, thus, is also static. The intent is that SOM monitoring would take place on a more frequent and spatially representative basis. Future indicators are likely to be developed by a more strategic monitoring of specific parameters such as SOM, soil acidity, salinity, and nutrient stocks, through a combination of remote sensing and community-level monitoring of variables of local interest.

The development of soil indicators is likely to demand and promote a two-way flow of information between communities and land management institutions. Researchers need to do much more work to develop indicators that relate soil quality changes to land use and management, and to their production, economic, and environmental impacts.

# WATER SERVICES

Water availability is an increasingly critical constraint to expanding food production in many of the world's agroecosystems. Agriculture accounts for the greatest proportion of withdrawals from the world's surface- and groundwater resources. And agriculture is the most consumptive user of water; that is, it returns the highest proportion of each cubic meter withdrawn to the atmosphere by evaporation and transpiration via plants. Thus, agriculture can profoundly affect the hydrological cycle of the watersheds in which it is practiced, and, consequently, the quantity and timing of available water resources. Additionally, and of growing concern, are the changes in water quality that agricultural production often entails. These range from increases in suspended solids—soil particles entering the surface water system through rainwater erosion of cultivated soils—to water pollution by the leaching of fertilizers, pesticides, and animal manure, to the accumulation of dissolved salts (salinization). These quantity and quality consequences of agricultural water use directly affect aquatic ecosystems and generate water-mediated impacts on a range of environmental, economic, and recreational goods and services as well as on human health.

Thus, we assess the relationship between agroecosystems and water-related goods and services in two ways. First, we consider water as a primary input to food, feed, and fiber production; second, we review the potential impacts of agriculture on water quantity and quality.

In keeping with the global agroecosystem characterization *(see section on Agricultural Extent and Agricultural Land Use Changes)*, the role of water in generating agricultural goods and services is divided into two distinct categories: rainfed systems, and extraction of ground- and surface water for irrigated systems.

## Condition of Rainfed Agroecosystems

Around 95 percent of all agricultural land, some 83 percent of cropland, depends on rainfall as the sole source of water. In these rainfed agroecosystems, the interaction of rainfall and evapotranspiration as mediated by crop and soil properties determines the availability of water for plant growth. As a measure of water availability, we used data from the FAO/IIASA Global Agroecological Zone database including temperature, rainfall, evapotranspiration, soil moisture holding capacity, and the length of growing period (LGP) for each year from 1960 to 1990 (FAO/IIASA 1999). LGP is the number of days per year in which moisture and temperature conditions will support plant growth. Map 4 shows the global variation in the average annual LGP within the PAGE global extent of agriculture, while Map 15 shows the year-to-year variability in length of growing period over the 30-year period. The regional patterns of growing season length and variability, both key factors in determining

*Table 21*

## Water Availability within the PAGE Agricultural Extent[a]

| | North America | Latin America and the Caribbean | Europe | Former Soviet Union | West Asia/ North Africa | Sub-Saharan Africa | South Asia | Southeast Asia | East Asia | Oceania | World |
|---|---|---|---|---|---|---|---|---|---|---|---|
| | | | | | Area Share by Region[b] | | | | | | |
| | | | | | *(percentage)* | | | | | | |
| **Rainfall (mm per year)[c]** | | | | | | | | | | | |
| 0-300 | 1.3 | 0.8 | 0.0 | 17.5 | 12.5 | 0.7 | 5.4 | 0.0 | 5.6 | 7.2 | 4.8 |
| 300 - 600 | 30.7 | 7.9 | 28.4 | 62.4 | 63.8 | 18.6 | 9.5 | 0.0 | 27.9 | 61.0 | 27.5 |
| 600 - 1000 | 35.0 | 24.9 | 56.6 | 19.7 | 23.5 | 37.0 | 37.2 | 0.0 | 19.7 | 19.8 | 29.2 |
| 1000 - 1500 | 31.8 | 32.8 | 12.3 | 0.3 | 0.1 | 33.3 | 32.0 | 17.8 | 30 2 | 3.9 | 23.0 |
| 1500 - 2000 | 1.1 | 24.9 | 2.3 | 0.0 | 0.0 | 7.4 | 8.5 | 22.5 | 15 3 | 3.8 | 9.5 |
| > 2000 | 0.0 | 8.7 | 0.4 | 0.0 | 0.0 | 2.9 | 7.4 | 59.7 | 1.4 | 4.1 | 6.1 |
| Region Total | *100.0* | *100.0* | *100.0* | *100.0* | *100.0* | *100.0* | *100.0* | *100.0* | *100 0* | *100.0* | *100.0* |
| **Length of Growing Period (days per year)[c]** | | | | | | | | | | | |
| Arid (0–74) | 20.8 | 5.3 | 0.0 | 29.3 | 12.9 | 6.3 | 10.9 | 0.0 | 5.5 | 14.6 | 11.7 |
| Dry Semiarid (75–119) | 6.1 | 5.2 | 0.5 | 14.7 | 30.2 | 12.3 | 15.2 | 0.0 | 9.9 | 21.6 | 9.8 |
| Moist Semiarid (120–179) | 12.0 | 13.4 | 13.3 | 32.3 | 37.3 | 33.3 | 47.6 | 0.4 | 19.0 | 29.9 | 23.2 |
| Subhumid (180–269) | 42.7 | 41.4 | 76.6 | 23.5 | 19.5 | 34.4 | 20.7 | 38.6 | 18.0 | 20.2 | 35.2 |
| Humid (270–365) | 18.4 | 34.7 | 9.6 | 0.1 | 0.0 | 13.8 | 5.6 | 61.0 | 47.7 | 13.7 | 20.1 |
| Region Total | *100.0* | *100.0* | *100.0* | *100.0* | *100.0* | *100.0* | *100.0* | *100.0* | *100.0* | *100.0* | *100.0* |
| **Year-to-Year Length of Growing Period Variability (coefficient of variation)[c]** | | | | | | | | | | | |
| 0 - 0 05 | 6.2 | 17.1 | 20.8 | 11.4 | 0.6 | 17.5 | 2.1 | 45.1 | 32 5 | 9.4 | 16 2 |
| 0.05 - 0.1 | 22.2 | 35.2 | 32.7 | 17.2 | 13.8 | 45.4 | 34.1 | 44.3 | 25 8 | 9.0 | 29 8 |
| 0.1 - 0.2 | 34.2 | 24.0 | 38.5 | 28.4 | 61.8 | 21.7 | 44.9 | 10.7 | 27.0 | 22.2 | 29 3 |
| 0 2 - 0.5 | 27.7 | 19.3 | 7.9 | 42.4 | 18.4 | 11.0 | 13.4 | 0.0 | 11 9 | 42.6 | 20 3 |
| 0 5 - 1.0 | 8.4 | 3.2 | 0.0 | 0.7 | 2.3 | 3.8 | 1.2 | 0.0 | 2.0 | 16.1 | 3.3 |
| > 1 0 | 1.3 | 1.3 | 0.0 | 0.0 | 3.1 | 0.6 | 4.2 | 0.0 | 0 7 | 0.7 | 1.1 |
| Region Total | *100.0* | *100.0* | *100.0* | *100.0* | *100.0* | *100.0* | *100.0* | *100.0* | *100.0* | *100.0* | *100.0* |

**Source:** IFPRI calculation based on: (a) GLCCD 1998; USGS EDC 1999. (b) ESRI 1996, and (c) FAO/IIASA 1999.
**Note:** These are not proposed as short- to medium-term indicators of agroecosystem condition. They are, however, good indicators of the rainfed production potential. Because they are relatively stable over time and not influenced directly by agroecosystem management decisions, they are useful as agroecosystem characterization or stratification variables *(see section on agroclimatic factors).*

agroecosystem output and management options, are summarized in Table 21. Regions with low and variable rainfall often support pastoral systems, while regions with high and stable rainfall favor high investments of labor and capital, improved pastures, and annual and permanent crops.

In rainfed agroecosystems, rainfall, LGP, and LGP variability are factors over which farmers have no control and that are not *directly* affected by agroecosystem management decisions. They are not, therefore, suitably responsive indicators of agroecosystem condition for the purposes of this study.[19] At a regional scale and over longer time periods, however, they do serve as intermediate indicators of the likely impacts of global climate change on agriculture (for example, Fischer et al. 1996), and as measures of the exposure of farmers to climatological risk. At a local scale, however, farmers do have management options for improving water availability (denser canopy, mulch-

ing, reduced tillage, soil conservation, and more water-efficient cropping systems).

## Condition of Irrigated Agroecosystems

Between 30 and 40 percent of the world's crop output comes from the 17 percent of the world's cropland that is irrigated, some 264 million hectares (WMO 1997:9; FAOSTAT 1999). This output includes nearly two thirds of the world's rice and wheat production. In India, for example, irrigated areas (one third of all cropland) account for more than 60 percent of total production (Rosegrant and Ringler 1997:10). Over the past 20 years, irrigated areas have steadily grown at approximately 1.5 to 2.0 percent per year (FAOSTAT 1999).

## EXTENT AND CHANGE IN IRRIGATED AREA

Expansion of irrigation not only signifies increasing water demands, but also implies more intensive land use with regard to other production inputs, such as fertilizers and pesticides. Extent can be expressed as an absolute area or as a share of total cropland. National time series data on irrigated area are available from FAOSTAT (1999). Two digital spatial sets containing irrigated land variables were available to this study: the University of Kassel Global Irrigation Area (Döll and Siebert 1999), and Global Land Cover Characterization Database (GLCCD 1998; USGS EDC 1999a). Both have shortcomings, but since the USGS data was particularly unreliable in detecting irrigated areas in South America, Africa, and Oceania, and the Kassel map was calibrated to FAO irrigation area statistics at the national level, the Kassel data source was preferred (see Map 16). Format and resolution issues precluded combining the data sets, but for comparative purposes the independently constructed maps were overlaid revealing significant areas of mismatch.[20] Given the importance of irrigated areas from an agricultural and environmental perspective, there is much to be done to improve our knowledge of their location and extent.

## IRRIGATION INTENSITY

There are many forms of irrigation at different scales of operation, but all share the objective of compensating for low rainfall by delivering sufficient water to the root zone to satisfy crop growth needs and, in some cases, to prevent the accumulation of dissolved salts. Crop water needs are large. Potato plants take up about 500 kg of water to produce 1 kg of potatoes, and in the cases of wheat, maize, and rice the corresponding requirements are 900 kg, 1400 kg, and 2000 kg of water respectively. Tropical crops, such as sugar cane and bananas, are even more demanding (Klohn and Appelgren 1998).

Of the 9,000 to 12,500 km³ of water estimated to be available globally for use each year (Postel et al. 1996:786; UN 1997), between 3,500 and 3,700 km³ was being extracted in 1995 (Shiklomanov 1996:69). Of that total, around 70 percent, some 2,450 to 2,700 km³ (WMO 1997:9; Postel 1993:58), is extracted for irrigation. High-income countries, mostly lying in the subhumid and humid temperate and subtropical regions, tend to have more abundant water resources—both per hectare and per capita. Poorer countries tend to have scarce water resources and relatively larger agricultural demands. According to the World Bank (2000:132), the share of extracted water used for agriculture ranges from 87 percent in low-income countries, through 74 percent in middle-income countries, to 30 percent in high- income countries.[21] So while the 206 m³ per capita withdrawn annually for agriculture by Africans represents 85 percent of their total withdrawals, the average of 1029 m³ per capita withdrawn by North Americans represents just 47 percent (WRI 2000: 276).

The simplest measure of irrigation intensity is the amount of irrigation water withdrawn (or applied) per year. This is most usefully expressed as an equivalent water depth, that is, cubic meters of water per year divided by hectares irrigated. For regional and global assessment, national estimates of the volume of freshwater extraction for agriculture are available from WRI (1998), and irrigated areas from FAOSTAT (1999). Using these data Seckler et al. (1998:32-38) calculated the depth of irrigation. Across all 118 countries covered, the mean depth of irrigation water extracted in 1990 was just under one meter—about the same as the average for China and a little less than the 1.1 meters for India (Seckler et al. 1998:32).

## IRRIGATION EFFICIENCY

Most irrigation systems use water inefficiently primarily because of the lack of incentives for farmers to treat water as a scarce resource. Efficiency is generally defined as the ratio of water actually used by crops (that is, returned to the atmosphere via transpiration) to the gross amount of water extracted for irrigation use. Extracted water is lost by direct evaporation from irrigation canals or the soil surface or by subsurface leakage. Global estimates of irrigation efficiency vary but average around 40 percent (Postel 1993:56; Seckler et al. 1998:25; Faurès 2000). At a watershed or basin level, however, these efficiencies can be misleading as a significant proportion of the subsurface leakage may be returned to surface- or groundwater resources elsewhere. If irrigation efficiency could be improved, less irrigation water would need to be extracted from rivers and aquifers per ton of food, feed, or fiber produced. Improved efficiency is being pursued in a number of ways: through policies that foster water markets or other regulatory arrangements; through technologies such as field leveling, low-energy precision application, drip irrigation, soil moisture monitoring; and through more water-efficient crops and cropping systems. There is also evidence of success through institutional reform and devolution that engage farming communities more directly in the improved management of water resources (Postel 1997; Subramanian et al. 1997).

Efficiency estimates were calculated based on country-specific, crop-related water use factors from Seckler et al. (1998) and FAO and WRI data were used for determining irrigation depth. Seckler et al.'s estimates suggest that arid agroecosystems have more efficient irrigation, e.g., 54 and 58 percent efficiency for the two driest groups of countries, compared to around 30 percent for the least water-constrained group. China and India show irrigation efficiencies of around 40 percent. They strongly influence the global average of around 43 percent (Seckler et al. 1998:25).

Table 22 summarizes the above irrigation indicators by region. Although there are many other potentially important indicators of irrigated agroecosystem condition, such as the pro-

*Table 22*

## Regional and Global Summary of Irrigation Indicators

| Region | Total Irrigated Area[a] | Area Growth Rate[a] | 1990 Irrigation Indicators | | | |
|---|---|---|---|---|---|---|
| | | | Included in the Seckler et al. Analysis | | | |
| | | | Irrigated Area[c] | Irrigation Abstraction[c] | Av. Irrigation Depth[b,d] | Av. Irrigation Efficiency[b] |
| | *(000 ha)* | *(%/year)* | *(000 ha)* | *(cubic km/yr)* | *(meters)* | *(percent)* |
| North America | 21,618 | 0.90 | 21,618 | 202 | 0.93 | 53 |
| Latin America and the Caribbean | 16,182 | 2.38 | 16,111 | 163 | 1.17 | 45 |
| Europe | 16,743 | 0.59 | 16,272 | 103 | 0.90 | 56 |
| West Asia/ North Africa | 22,570 | 2.45 | 21,805 | 219 | 1.17 | 60 |
| Sub-Saharan Africa | 4,773 | 1.20 | 4,604 | 53 | 1.59 | 50 |
| South Africa | 1,290 | 0.77 | 1,290 | 15 | 1.16 | 45 |
| Asia | 154,449 | 1.87 | 136,564 | 1,324 | 1.02 | 39 |
| India | 45,144 | 2.73 | 45,144 | 484 | 1.07 | 40 |
| China | 47,965 | 1.31 | 47,965 | 463 | 0.97 | 39 |
| Rest of Asia | 61,340 | 1.70 | 43,455 | 377 | 0.92 | 32 |
| Oceania | 2,113 | 3.57 | 2,112 | 6 | 0.29 | 66 |
| World | 243,028 | 1.57 | 220,376 | 2,086 | 0.95 | 43 |

**Source:** (a) compiled from FAOSTAT 1999. (b) IFPRI calculations based on Seckler et al. (1998:32-38). (c) Seckler et al. (1998:32-38). (d) Gross equivalent depth-abstraction divided by area. Seckler et al. provide summaries for countries grouped by water scarcity. Total global irrigation abstraction in 1990 is estimated at 2,353 cubic kilometers (Seckler et al. 1998:32).

ductivity of irrigated areas per unit of irrigated land, cropped area, or irrigation water, the additional crop-specific data they require are usually unavailable at a macro level (see Molden et al. 1998 for a discussion of other relevant indicators).

## SALINIZATION AND WATERLOGGING

Two significant environmental consequences of irrigation are salinization and waterlogging. Salinization occurs through the accumulation of salts deposited when water is evaporated from the upper layers of the soil. Although salinization can occur naturally, irrigation promotes so-called "secondary" salinization because it artificially increases the supply of water to surface layers of the soil in typically more arid climates where evaporation rates are higher. This accelerates the build-up of salts. Because most crops are not tolerant of high levels of salinity, salinization of land and water can in extreme cases lead to the abandonment of irrigated areas, but is more usually reflected in declining yields. The problem is particularly acute in arid and semiarid areas, such as in Pakistan and Australia, where soil evaporation rates are much higher and natural leaching and drainage are inhibited. In the case of Australia, salinization occurs chiefly in rainfed areas where natural water tables have been rising since settlers first began clearing the natural bush vegetation, primarily for grazing land.

Waterlogging is more prevalent in humid environments and in irrigated areas where excessive amounts of water are applied to the land and the water table rises. Waterlogging is often a precursor to salinization. Both salinization and waterlogging most often arise from poor irrigation management and inadequate drainage. A review of estimates of land damaged by salinization put the global total at around 45Mha, representing some 20 percent of the world's total irrigated land (Ghassemi et al. 1995:42). Rough estimates of the annual impacts of degradation in irrigated areas, primarily through salinization, are losses of around 1.5Mha of irrigated land in the world's dry areas (Ghassemi 1995:41 quoting Dregne et al. 1991), and approximately US$11 billion from reduced productivity globally (Postel 1999:92). These losses represent just under 1 percent of the global totals of both irrigated area and annual value of production, but are much more significant in affected areas.

## Agricultural Effects on Water Quality

Both rainfed and irrigated agriculture can markedly affect the quality of water in ways other than salinization. Poor crop cover, field drainage, and cultivation operations, particularly on sloping land, can lead to increased levels of water-induced soil erosion. Generally, increases in water-borne suspended solids negatively affect downstream aquatic ecosystems, as well as cause siltation on downstream channels, reservoirs, and other hydraulic infrastructure. Analysts estimate that around 22 percent of the annual storage capacity lost through siltation of reservoirs in the United States is due to soil erosion from cropland (Gleick 1993:367). But the linkages between land use practices, erosion, and sedimentation in rivers are complex. Box 3 summarizes some recent findings on this issue.

High external-input systems are prone to generate significant environmental externalities—typically, pollution from fertilizer, pesticides, and animal manure leaching into ground- and surface water. In Belgium and the Netherlands, the nitrogen input to some croplands exceeds 500 kg/ha. Years of phosphate fertilizer application in parts of the United States has left many soils saturated with phosphorus (FAO 1999b:172). Leaching of excess nutrients from farms into water sources causes eutrophication, which consequently damages aquatic plant life and fauna through algal blooms, depressed oxygen levels, and increased water treatment costs. In addition, excess nitrates pose direct human health threats. The depletion of river flows exacerbates such water quality problems by reducing the dilution capacity of rivers and downstream water bodies. Box 4 describes an example of an agricultural water pollution controversy in the United States, and the difficulties faced in dealing with it, even when environmental regulations and enforcement measures are well established.

Monitoring water for nutrients, ions (salinity), solids, persistant organochlorine pollutants (POPs—originating in pesticides and related compounds), and others, remains the most reliable means of tracing changes in water quality. But even though cheaper, simpler means of field testing are continually being developed, water quality monitoring remains technically and financially demanding. For example, it is necessary to correct some observations for water temperature, pH, and flow rate effects. Even with reliable water quality data, it is often difficult, especially in watersheds having significant human populations and commercial activities, to clearly relate changes in such quality indicators to the (generally diffuse) effects of agricultural activity. For example, there is recent evidence from the United States of higher prevalence of pesticides and nutrients in surface runoff from urban than from agricultural catchments while, in the case of groundwater, prevalence is higher in agricultural catchments (USGS 1999).

## Agricultural Effects on Water Supply

With regard to land use change, there is considerable evidence that converting forest and woodlands to agricultural uses increases available surface- and groundwater resources because of the substantial reduction in biomass and, consequently, in transpiration demands (FAO 1999b:173). However, reduced biomass and litter, coupled with intentional efforts to improve drainage, reduce the water storage capacity of agricultural land and can seriously diminish its flow regulation capacity. This increases the incidence and severity of high runoff events and, correspondingly, diminishes dry weather flows.

*Box 3*

### Historical Decline in Soil Erosion in U.S. River Basin

One of the most intensive, longitudinal studies of soil erosion in the world was recently completed for the Coon Creek Basin, located in the humid midwestern United States. The Basin drains into the Mississippi River. The study documented changing sedimentation rates in a 360-square kilometer area, comparing original prairie soils present when European farmers arrived in the 1850s with detailed erosion studies by the U.S. Soil Conservation Service in the 1930's and resurveys of soil profiles in the 1970s and the 1990s.

Measured rates of sedimentation jumped in the late nineteenth century, skyrocketed in the 1920s and 1930s, and then dropped again as United States Department of Agriculture pressed local farmers to stop using the traditional moldboard plow and adopt conservation practices. These practices included strip-cropping and leaving plant residue and stubble in the fields year-round to inhibit run-off. Between the 1970s and the 1990s, sedimentation rates dropped to just 6 percent of their previous peak. The findings illustrate the potential for conservation practices, when widely adopted, to dramatically reduce soil degradation.

Yet, the study also found that regardless of erosion rates, the Basin tended to store and release sediment in such a way that the amount delivered to the Mississippi River remained roughly constant over the decades. Sediment eroded from upland areas is, in effect, stored around Coon Creek tributaries and other deposition sites and released later. For example, cutbanks and floodplains around the oxbows of the tributaries changed in shape and size over time and were transformed from sediment sources to sinks and back again. The findings suggest that even if materials coming off a field or group of fields are controlled, it may be some time before effects on downstream sedimentation are observed. Thus, short-term measures of downstream sedimentation may be a poor indicator of the quality of upstream farmers' land husbandry practices.

**Source:** Adapted from Trimble, S.W., *Science*, vol. 285, 20 August 1999, pp. 1187-89.

Decreased river flows and falling groundwater levels are pervasive in irrigated areas, because few incentives exist to not overuse water. In the United States, roughly one fifth of the irrigated area (about 4 million hectares) is estimated to be extracting groundwater at greater than the recharge rate (Postel 1993:58). Postel (1993:59) and Rosegrant and Ringler (1997:419) report water tables in the North China Plain falling by up to one meter per year, and by 25-30 meters in a decade in parts of Tamil Nadu, India. A notorious case of river water

Box 4

# Cheaper Chicken, More Pollution

Between 1970 and 1996, per capita consumption of poultry meat in the United States nearly doubled. Health concerns and demographic changes stimulated demand, while increased retail competition and new chicken production techniques were responsible for the increased supply, which drove down chicken prices *(see Figure A)*. To reduce the costs of transporting feed and chickens, large, vertically-integrated producers contract with nearby farmers to raise chickens for company slaughterhouses. As more farmers enter the industry, the concentration of chickens and slaughterhouses increases, and so does the problem of manure disposal.

The chicken industry in Delmarva (the peninsula of Delaware, Maryland, and Virginia) exemplifies the problem. The area produces more than 600 million birds a year and 750,000 tons of manure—more than that produced by a city of 4 million people. Farmers have traditionally used chicken manure, which is richer in nutrients—especially nitrogen (nitrates) and phosphorous—than other livestock waste, to fertilize their fields. But, as housing development expanded into farmland, more manure was applied to less cropland and soils became saturated with nutrients. A survey conducted in the late 1980s found that one third of all groundwater in Delmarva's agricultural areas was contaminated with nitrates, confirming that excess nutrients were seeping into and polluting groundwater, wells, and other public water sources.

At the same time, company slaughterhouses disposed of millions of gallons of wastewater each day in waterways that reach the Chesapeake and coastal bays. Other creeks and streams carry surface run-off from overfertilized fields or piles of wet manure. Although wastewater should be cleaned to per-unit standards, its growing volume raises the total load of nutrients beyond acceptable levels. Treating slaughterhouse wastewater also creates sludge (minute solids filtered from the wastewater) that is injected into the ground for disposal, increasing groundwater pollution.

The Environmental Protection Agency Chesapeake Bay Program estimates that chicken manure sends more than four times as much nitrogen into the Bay as the biggest nonagricultural sources (septic tanks and run-off), and five times as much phosphorous as sewage-treatment plants. Storms exacerbate the problem, washing large amounts of nutrient-laden sediment into the Bay. The excess nitrogen and phosphorous overstimulate algae growth. As the algae die and decompose, they consume oxygen, choking fish and other water life *(see Figure B)*. Some scientists believe the toxic microbe *Pfiesteria piscicida* feeds on the excess algae and nutrients.

So far there is no agreement on who should clean up the pollution the industry produces or which methods should be used. Some companies have begun trucking manure to more distant farming areas. Others are exploring technological solutions, including adding phytase to chicken feed to help the

**Figure A**

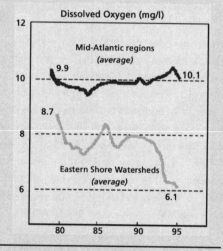

**Source:** USDA/ERS, Poultry Yearbook 1996; Data downloaded from http://usda.mannlib.cornell.edu/

**Figure B**

**Source:** The Washington Post.

chickens digest phosphorous and building plants to convert manure into fuel and fertilizer pellets.

Government regulation has so far failed to solve the problem. The industry's economic importance makes it difficult for local politicians to pass and enforce more stringent antipollution legislation. One study estimated that a 4 percent drop in the state's poultry production would eliminate 1,000 jobs and $74 million in economic output. In 1997, the EPA fined one company $6 million after state regulators failed to stop the discharge of excessive nutrients. The problem has pitted environmentalists, federal and state regulators, and the industry against each other. Industry leaders complain that they are caught between the demand for cheap chicken and the desire for pristine waterways.

**Source:** Adapted from Washington Post series *Poultry's Price: The Cost of the Bay.* August 1-3, 1999.

overextraction for irrigation is the remarkable shrinkage and salinization of the Aral Sea and the consequent loss of fish species and fishing livelihoods (Gleick 1993:5-6; Postel 1993:59).

For irrigated systems, the indicators of area, depth, and efficiency described earlier provide a broad picture of water resource use. To place that extraction in context, however, requires estimates of the renewable capacity of water resources. Only by comparing existing and likely future demands against renewable resource capacity will it become apparent how water scarce each location is, what stresses agroecosystems and other potential water users face, and what options might be available to balance demands with renewable supplies, including, for example, increased water reuse. Assessing the resource potential for surface water is relatively straightforward compared to assessing reliable groundwater yields, in many places the major source of irrigation water. This study has not attempted to assess the reliable yields of water resources.

Given these complexities and the limited scope of this study, we adopted a much simpler indicator, the proportion of a basin within the PAGE global extent of agriculture, as a guide to the potential impact of agroecosystems on water goods and services. This indicator is shown at the global scale in Map 17. The indicator is continuous, but is displayed here in just three ranges (10-30 percent, 30-50 percent, and greater than 50 percent) to simplify map presentation. Because of its simplicity, the indicator can easily be updated as new land cover or watershed boundary data become available. The map depicts the area extent of agriculture within each watershed boundary. Given that one hectare of irrigated area will likely have a much greater impact on water goods and services than one hectare of rainfed agriculture, the indicator could be further refined by differentiating between rainfed and irrigated agricultural areas within each watershed. Such a differentiation was not made here because of resolution differences among the satellite-derived agricultural extent data, the irrigation area data, and the basin boundaries.

## Summary of Indicators and Data

Water has a central role within and beyond agriculture. The need to use those finite resources more effectively in the face of growing demands and greater pollution threats has already mobilized several international initiatives, such as the Global Water Partnership and the World Water Council. The most recent manifestation of the international concern was the Second World Water Forum held in The Hague, the Netherlands, in March 2000 (World Water Forum 2000). One relevant outcome from that meeting was the commitment by the UN systems to produce periodic reports on the state of the world's water resources, a so-called "water development report."

An important objective of all such initiatives is to overcome the lack of reliable and internationally comparable water information, a constraint reflected in several parts of this chapter. The proposed Millennium Ecosystem Assessment should link with and add value to the networks and resources of ongoing programs including: the Global Climate Observation System (GCOS 2000); the World Hydrological Cycle Observing System (WYCOS 2000); the Global Environment Monitoring System for Water (GEMS/Water 2000); and the Global Runoff Data Centre (GRDC 2000). Some of these resources are significant. For example, the GEMS/Water archive contains over 1.5 million data points for 710 monitoring sites globally collected since the late 1970s. Although the sampling sites are more numerous in developed countries, the standardized data sets do include several of the key nutrient, ion, solids, and organic contaminant indicators discussed above.

With specific reference to water and agriculture, the FAO/Netherlands Conference on the MultiFunctional Character of Agriculture and Land (MFCAL) held in Maastricht, the Netherlands, in September 1999 explicitly addressed the multiple goods and services of water in similar ways to those explored in this study (FAO 1999b). FAO is also responsible for the AQUASTAT initiative established specifically to compile and disseminate water information, with a strong focus on improving irrigation and drainage data (FAO 2000a). The University of Kassel compiled the irrigation map used in this study (Döll and Siebert 1999) and is continuing to refine that map (in collaboration with FAO's AQUASTAT program). In addition, it is testing global water use scenarios for the World Water Commission (Alacamo et al. 2000). This scenario development activity also involved Stockholm Environment Institute (SEI) (Gallopin and Rijsberman 1999) and IWMI (IWMI 1999) who focused on the agricultural dimensions, as well as the International Food Policy Research Institute (IFPRI) who incorporated the water scenarios into global food assessments to trace the likely consequences on food availability and prices (Rosegrant and Ringler 2000).

Field level indicators of the efficiency of rainfall use under different management practices would be valuable to local agroecosystem assessments. Satellite-derived data on rain-use efficiency (RUE) is now available (University of Maryland 1999; Prince et al. 1998) and its suitability as a regional-scale indicator of rainfed agroecosystem condition merits further investigation. A companion PAGE report, *Pilot Analysis of Global Ecosystems: Grasslands* (White et al. 2000), reviews the use of RUE as a condition indicator. Additionally, the PAGE report, *Pilot Analysis of Global Ecosystems: Freshwater* (Revenga et al. 2000) reviews and analyzes the state of the world's freshwater ecosystems.

# BIODIVERSITY

Before crops were first domesticated some 10,000 years ago, humans met their nutritional and other welfare needs by hunting and gathering within natural ecosystems. Since then, agriculture—the purposeful selection and domestication of valuable plant and animal species—has increasingly met these needs by producing domesticated species on land formerly occupied by forests, shrubs, grasses, mangroves, and other ecosystems. Thus, the conversion of natural to agricultural ecosystems has always involved change in the type, mix, and function of plant and animal species. But the global expansion of agriculture to meet growing food needs and the intensification of production systems have had a profound, and largely negative, effect on natural and domesticated plant and animal biodiversity.

There are three key, interlinked factors underlying agriculture's threat to biodiversity: population growth, changes in agricultural productivity, and production practices. The relationship between population growth and productivity growth conditions the aggregate demand for agricultural land, while production practices condition the scope for biodiversity to thrive, or otherwise, within agricultural land. Our PAGE assessment recognizes four ways in which agriculture is currently affecting biodiversity. First is the large-scale conversion of land to agroecosystems, with the consequent loss of natural habitats. Second is the composition and spatial structure of agricultural landscapes that can significantly reduce their habitat value. Third is the loss of wild species as a direct consequence of agricultural inputs and practices, such as the toxic effects of some pesticides on birdlife. Fourth is the general loss of diversity among and within the economic plant and animal species grown in agricultural systems.

Some of these losses impair the performance of agriculture itself, by affecting above- and below-ground biological systems that play a critical role in pollination, control of agricultural pests, breakdown of agricultural residues and wastes, and recycling nutrients critical to plant growth (Swift and Anderson 1999). For example, populations of about 1,200 wild vertebrate pollinators are so reduced that they are listed as endangered species. Their decline is depressing yields of blueberries and cherries in Canada, cashew nuts in Borneo, Brazil nuts in South America, and pumpkins in the United States (McNeely and Scherr forthcoming).

The presence of wild species in agroecosystems is also important in many regions because of the limited potential to establish natural biodiversity reserves. In such areas, and else-

*Table 23*

## Share of Major Habitats[a] Occupied by Agriculture[b]

| Major Habitat Type[a,c] | Share Occupied by Agriculture[b,d] |
|---|---|
| Tropical and Subtropical Conifer Forests | 25.2 |
| Tropical and Subtropical Dry and Monsoon Broadleaf Forests | 43.4 |
| Tropical and Subtropical Grasslands, Savannas, and Shrubland | 17.9 |
| Tropical and Subtropical Moist Broadleaf Forests | 19.2 |
| Temperate Broadleaf and Mixed Forests | 45.8 |
| Temperate Conifer Forests | 16.4 |
| Temperate Grasslands, Savannas, and Shrublands | 34.2 |
| Montane Grasslands and Shrublands | 9.8 |
| Flooded Grasslands and Savannas | 20.2 |
| Deserts and Xeric Shrublands | 30.2 |
| Mediterranean Shrublands | 4.5 |
| Boreal Forests/Taiga | 7.2 |
| Tundra | 0.9 |
| Mangroves | 26.8 |
| Islands | 0.6 |
| Glaciers and Rocks | 0.3 |

**Source:** IFPRI calculation based on: (a) the World Wildlife Fund's (WWF-US) Ecoregions Data Base (Olson et al. 1999), and (b) the PAGE Agricultural Extent (reinterpretation of GLCCD 1998; USGS EDC 1999).
**Notes:** (c) An ecoregion is defined as a relatively large area of land or water that is characterized by a geographically distinct set of natural communities that (1) share a majority of their species, ecological dynamics, and environmental conditions, and (2) function together as a conservation unit at global and regional scales. The ecoregions were grouped into major habitat types (MHTs), which are defined by the dynamics of the ecological systems and the broad vegetative structures and patterns of species diversity within them (Olson et al. 1999 cited in Matthews et al. 2000). (d) The area within a habitat type that is agricultural was determined by applying a weighted percentage to each PAGE agricultural land cover class (80 percent for areas with at least 60 percent agriculture; 50 percent for areas with 40-60 percent agriculture; 35 percent for areas with 30-40 percent agriculture; and 5 percent for areas with 0-30 percent agriculture).

where, agroecosystems are under increasing pressure to play a much greater role in providing habitats, while continuing to provide food and other goods and services.

The following sections describe indicators developed to assess the status of agroecosystems in the context of the four biodiversity impacts identified above.

## Status of Land Conversion to Agricultural Use

Globally, agricultural land has expanded by around 130,000 km² year over the past 20 years (FAO 1997b:17), predominantly at the expense of natural forests and grassland, the major repositories of the world's diversity of plant and animal species. This is a net increase, representing the amount by which land conversion into agriculture exceeds conversions out of agriculture. Net area expansion, therefore, understates the true scale of agricultural conversion impacts on habitat and associated biodiversity. For comparative purposes, tropical deforestation rates range from about 50,000 to 170,000 km² per year (Tucker and Townshed 2000:1461-1472). Given that land is also con-

verted into agriculture from other sources (*see Box 1 for the case of China*), a significant proportion of forest and grassland conversion must also be attributable to nonagricultural pressures, such as logging and urban expansion.

The two indicators, described below, have been developed to assess the extent to which natural habitats have been converted to agriculture, and the extent to which agriculture may be threatening habitat reserves.

### CONVERSION OF NATURAL HABITATS TO AGRICULTURE

The greatest habitat disturbance, within the PAGE defined extent of agriculture, is likely to be found in agriculture-dominant areas (>60 percent agriculture), although areas with moderate influence (30-60 percent agriculture) may also experience significant disturbance to wild species mix and numbers. The extent to which different habitat types have been converted to agriculture is quite variable. Table 23 summarizes the proportion of each major global habitat type that agriculture occupies, based on the World Wildlife Fund – U.S. (WWF-US) Ecoregions Database (Olson et al. 1999). The habitats most radically affected are temperate broadleaf and mixed forests, and tropical and subtropical dry and monsoon broadleaf forests, of which nearly half the global area has been converted to agricultural use. Although the *rates* of tropical rain-forest clearing are high, only a fifth of all tropical and subtropical moist broadleaf forests appear to have been converted to agricultural use. This finding, however, is likely underestimated, as a result of problems in using satellite data to detect agricultural land uses in and around abundant natural tree cover, or to distinguish rotational fallow from undisturbed forest. "Hot spots" for habitat conversion include the following: the forest frontiers of Indonesia, Malaysia, Vietnam, Cambodia, and Laos; protected areas in Madagascar; the Central American and Amazon rain forests; the Pacific rain forests of Colombia and Ecuador; the Chaco region of South America; and the Atlantic lowlands of Central America (Scherr and Yadav 1995).

## AGRICULTURE IN HABITAT RESERVES

Over the past century, and particularly in the past few decades, national governments around the world have sought to preserve natural biodiversity and habitats by establishing "protected areas." In these areas, agricultural use is generally legally prohibited, limited, or regulated. The indicator selected as a measure of the pressure on protected areas from agriculture, as well as of the capacity to protect ecosystems within agricultural areas, was the extent to which agriculture is practiced in protected areas (as an area percentage). This is a continuous measure that can be used to monitor the success of biodiversity protection and conservation policies. At the global scale, we constructed the indicator using the protected area spatial database from the World Conservation Monitoring Centre (WCMC 1999), and the PAGE agroecosystem extent map.

It is evident from Map 18 that protected areas have been widely established in and around agricultural lands, as defined by the PAGE agroecosystem extent, with the exception of North America (although there are many farm wetland conservation sites in the United States unmapped by our data), and the Former Soviet Union. In Central America, many small, protected areas are interspersed in agricultural land. Because of limited total land area, agricultural pressures on protected areas such as these are likely to be intense. Many protected areas in South America are located along the agricultural frontier of the Amazon and their protection depends greatly on the dynamics of frontier settlement and employment opportunities in other areas.

In Europe and southwest Australia, many protected areas are under pressure from agricultural use. More effective institutions and regulations, higher incomes, public support for biodiversity conservation, and declining employment in agriculture make it more likely that protected areas can be conserved despite this spatial pattern. By contrast, in Africa, South, Southeast, and East Asia there is a striking overlap of protected areas with the agricultural extent. The challenge to protect these areas effectively seems daunting.

## Status of Natural Habitat within Agricultural Landscapes

Crop, livestock, and tree crop production typically cover only a portion of agricultural landscapes, even in intensive systems. Theoretically, much of the remainder could serve as habitat where native plant and animal species could flourish. However, elements of the landscape structure strongly affect its suitability as habitat, including: patch size of natural areas (determined by land use fragmentation); the presence of barriers to species movement, such as roads and irrigation canals; the presence of refuges and buffers around cropped areas, and the presence of "corridors" that connect patches of habitat, such as streambank vegetation. Given the lack of information to describe agricultural landscapes in this way, even at a local level, there are relatively few current options for representing natural habitat quality in agroecosystems. This section presents the single indicator estimated for the PAGE study—tree cover—and briefly describes two more with potential for further development.

### TREE COVER IN AGRICULTURAL LANDS

Farmers grow trees in agricultural areas for a variety of economic, social, and cultural reasons. Trees also provide key environmental services to farmers by stabilizing soil and runoff and are often found in ecologically strategic niches along streambanks or protecting steep slopes. In addition to the greater biodiversity the trees themselves represent, trees are habitat for other flora and fauna, birds, insects, and soil microorganisms in particular. Thus, we identify the proportional area of trees found within the agricultural extent as a proxy indicator for the availability of natural habitat within agricultural landscapes.

Researchers from the University of Maryland have interpreted the same detailed satellite data that underpins the land cover classification used by PAGE to derive additional vegetation characteristics including woody vegetation, defined as mature vegetation whose approximate height is greater than 5 meters (DeFries et al. 2000). The tree cover database was combined with the PAGE agricultural extent map to produce Map 19.

The map shows major regional differences in the proportion of tree cover within the PAGE agricultural extent. The proportion of tree cover is lowest in the agricultural lands of West Asia/North Africa, which are long-settled and dry, with limited old-growth forest cover, and dominated by livestock and cereals. Cover is also relatively low in Australia and the islands of Oceania, probably because of dryer conditions, lower initial forest cover, and agricultural land scarcity. In North America,

Figure 13

## Production System Classification by Diversity

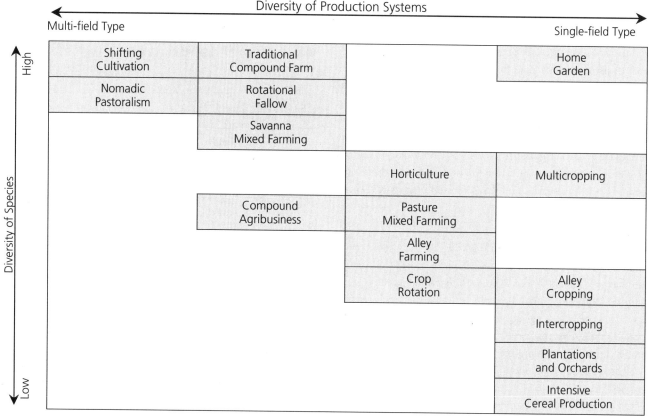

**Source:** Swift and Anderson 1999:20.

Europe, Former Soviet Union, and East Asia, over half of the agricultural extent has little or no tree cover, despite high original extent of natural forests, probably because of patterns of agricultural land-clearing and mechanized operations, and, in some cases, land scarcity.

In Latin America, Sub-Saharan Africa, South Asia, and Southeast Asia, a majority of the agricultural extent has significant tree cover, except in the intensively cultivated irrigated lands and some drylands in South Asia. In these regions, tree and shrub fallows are still common in less densely populated areas, as are tree crop plantations, such as tea, coffee, bananas, and oil palm in tropical areas. Additionally, farm households depend more on locally grown trees for fuel, construction, raw materials, fencing, and forage. Because agricultural landscapes are more variable, there are also more microniches on farms that might be better suited to trees than to crops.

### FRAGMENTATION

A major impact of agricultural expansion and intensification has been the fragmentation of habitats, which alters vegetation, nutrient, water, and microclimatic regimes. The construction of roads and other infrastructure in densely populated areas further dissects the landscape, degrades vegetation, and limits movement of wildlife and dissemination of plant seeds. Indexes of fragmentation derived from remotely sensed images have been developed for many forest areas (Matthews et al. 2000; Smith et al. 2000:28-32) but modifications are needed for their application to agricultural lands. The further development of fragmentation indicators in agricultural landscapes will significantly contribute to a better understanding of the biodiversity impacts of agriculture.

### FARMING SYSTEMS DIVERSITY

Farmer decisions on cropping systems include crop and variety selection, as well as the configuration of production spatially (monocropping, intercropping, alley cropping) and in time (crop rotation or relay cropping). At the landscape level, within and across farms, the use of multiple varieties can spread production and market risk, stagger maturity dates, cater to different uses, and take advantage of different microenvironments within the farm extent. For, example, large farmers or farming communities may adopt planting patterns and varietal mixes that limit the scope for pest and disease outbreaks. Production systems decisions do, thus, significantly affect the availability and quality of habitat for wild biodiversity.

Figure 13 characterizes production systems in terms of their biological and spatial diversity. Polycultures (usually with perennial components) are species diverse but are relatively stable over time at the variety level. Monoculture (single species) systems, by contrast, have higher varietal turnover, because they are generally more susceptible to pests and diseases and are better served by scientific research. Polycultures usually offer a more benign environment for a broad range of plant, soil biota, insect, and other animal species to flourish. It seems worthwhile to develop indicators of production system diversity not only as a proxy for the potential quantity and quality of habitat within the cultivated areas of agroecosystems, but also for the insights it would provide on the diversity of agricultural species.

## Status of Wild Biodiversity in Agroecosystems

Effective global monitoring of wild biodiversity in agroecosystems requires more than indicators of potential habitat value. Data must be collected, at some point, on the actual presence of wild species. Many case studies of species richness and diversity within local farming systems are now available, and new, lower-cost means of assessment are being developed (for example, Daily et al. 2000; Ricketts et al. 2000). Currently, there are no aggregate data on species numbers or populations that could be used as global indicators.

Indirect approaches could be developed in the future. Remote sensing methods could be used to assess soil organic matter, as a proxy for the level of biota in the soil *(see Soil Resource Condition section)*. Another approach is to monitor "density of indicator species" whose presence is known to be associated with a relatively healthy habitat for a range of other wild species, such as a bird known to be sensitive to agrochemical pollution. Indicator species would need to be easily observable and would ideally be species which farmers might have an incentive to monitor, for example, beneficial carnivore birds, such as hawks and owls, that keep down crop, insect, and rodent pests, or valuable crop pollinating insects.

## Status of Agricultural Diversity

Notwithstanding debates on biotechnology, diversity among and within domesticated crop, livestock, and tree crop species currently constitutes the genetic core that sustains agriculture. Many view the genetic resources developed and maintained over centuries as a heritage of present and future generations. In the pursuit of improving crops and livestock, agricultural breeders seek to maintain the diversity of domesticated species, as well as their wild relatives, as potentially important sources of desirable genetic traits.

For the PAGE study, we propose three indicators of agricultural diversity: information on species level diversity, varietal diversity, and the uptake of transgenic crops. It was not possible within PAGE to satisfactorily develop these indicators on a global basis, but several major international efforts are currently addressing the significant information challenges involved: the World Information and Early Warning System on Plant Genetic Resources – WIEWS (2000), the International Crop Information System – ICIS (2000); and the Domestic Animal Diversification Information System – DAD-IS (2000).

### AGRICULTURAL SPECIES DIVERSITY

In subsistence-oriented production systems, farmers must grow a wide range of crops to meet nutritional and other needs. Some production systems, such as home gardens and intercropped agroforestry systems, can achieve high levels of both agricultural and wild biodiversity. However, as markets become more developed, factors such as lower transaction costs, comparative advantage in production and processing, economies of scale, and increased labor opportunities outside agriculture, favor the emergence of intensive, specialized cropping systems. This specialization has reduced the diversity of agricultural species grown globally and within individual farmers' fields (Thrupp 1998).

*Crop biodiversity*. Of the 7,000 crop species believed to have been used in agriculture (only a portion of the 10,000-50,000 estimated number of edible plants), less than 2 percent are currently important at the national level, of which only 30 provide an estimated 90 percent of the world's calorie intake with wheat, rice, and maize alone providing more than half of the global plant-derived calories (FAO 1998:14; Shand 1997:23).

*Livestock biodiversity*. Of the approximately 15,000 species of mammals and birds (IUCN 1996), some 30-40 (0.25 percent) have been used extensively for agriculture, with fewer than 14 accounting for over 90 percent of global livestock production (UNEP 1995:129). Few new animal species have been domesticated through modern science. But animal domestication over the last 10,000 years has led to numerous breeds of these species and, until recently, greater genetic diversity than is normally found within a wild species. With the current trends of homogenization, FAO estimates that in Europe 50 percent of livestock breeds that existed 100 years ago have disappeared. Some 30 percent of domesticated breeds found internationally are at risk of extinction (Shand 1997:46).

*Tree crop biodiversity*. With the exception of some tree species producing highly valued fruits, nuts, oils, or internationally traded beverages, the domestication of tree crops has lagged considerably behind that of annual crops and livestock. Although tens of thousands of wild tree and shrub species have probably

*Box 5*

## Potato Biodiversity in the Andes

Farmers in the Andes of South America cultivate hundreds of native varieties belonging to diverse species of potato. In this region, people first domesticated this crop several thousand years ago. But genetic diversity is not uniformly distributed throughout the region *(see Panel a)*. In the Northern Andes of Colombia and Ecuador, potatoes are increasingly produced in specialized compact regions *(see Panel b)*. Potatoes can be planted year-round, and drought stress does not substantially limit yield potential. Only two of the eight cultivated species of potatoes are planted in Colombia. Since the 1960s, the leading variety in Colombia, Parda Pastusa, has accounted for more than 50 percent of growing area.

In contrast, in the Southern Andes, potato production is more subsistence-oriented and geographically more dispersed *(see Panel b)*. Here, there is a short growing season with drought and frost adversely affecting yield potential. Farmers in Peru and Bolivia cultivate up to eight species *(see Panel a)*. A native variety, Waych'a, is still the most widely grown clone in Bolivia, but it only accounts for 10-20 percent of area. As many as 180 locally named varieties have been encountered in some Bolivian communities.

The yields in the Northern Andes are about three times higher than in the Southern Andes. Intensive potato production in the Northern Andes does not appear to negatively affect the environment. Field work in Carchi, the main potato producing region in Ecuador, has shown, however, that agricultural pesticides cause serious dermatological and neurobehavioral problems for farm workers (Crissman et al. 1998). Improving worker safety and enhancing the adoption of integrated pest management (IPM) for the Andean potato weevil could reduce the adverse health effects of pesticide use by as much as 50 percent without a significant impact on production.

With increasing economic integration in the Andes, the yield-potential advantage of the Northern Andes should translate into a gradual increase of potato exports to the Southern Andean countries. Nevertheless, the threat of increasing commercialization to the genetic diversity of the crop is remote because modern potato varieties are generally not well adapted to the harsher conditions in the Southern Andes and consumption preferences are strong for native varieties. Moreover, disease-resistant varieties recently released by the national potato program in Bolivia will lead to increased temporal diversity of this important food crop.

*Panel a*

### Species diversity in cultivated potato in the Andes

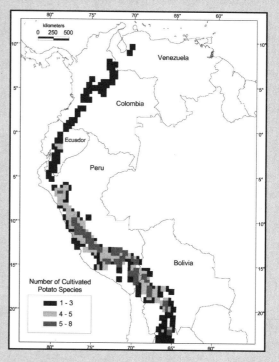

*Panel b*

### Altitude and distribution of potato growing area in the Andes

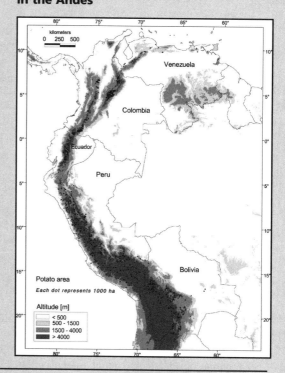

---

*Source*: Hijmans, R. 1999. Potato Biodiversity in the Andes. Special report prepared for IFPRI as part of the PAGE study. Lima, Peru: Centro Internacional de la Papa.

**Table 24**

## Adoption of Modern Rice and Wheat Varieties in the Developing World by Moisture Regime, Mid-1980s

| Modern Rice Varieties | | | Modern Wheat Varieties | | |
|---|---|---|---|---|---|
| Environment | Production[a] | Area[b] | Environment | Production[a] | Area[b] |
| | (percentage) | | | (percentage) | |
| Irrigated lowlands | 71 | 95 | Irrigated | 49 | 91 |
| Rainfed lowlands | 19 | 40 | Rainfed (>500mm)[c] | 28 | 60 |
| Deepwater | 7 | 0 | Rainfed (300 500mm)[c] | 22 | 45 |
| Upland | 4 | 0 | Rainfed (<300)[c] | 1 | 21 |
| Total | 100 | 48 | Total | 100 | 62 |

**Source:** Byerlee 1996:699.
**Notes:** (a) Production column shows the share of total production in each environment. (b) Area column shows the proportional area of each environment in which modern varieties have been adopted. (c) Rainfall immediately prior to and during the growing season.

been used for one purpose or another in diverse parts of the world, a much smaller number are intentionally grown and managed on farms. Seeds are currently commercially available for 2,632 tree species (Kindt et al. 1997); an unspecified number are propagated vegetatively from wild plants. The slower commercial development of forest trees historically has been due in part to the abundance of naturally growing trees in less intensive agricultural systems, and limited markets to sell surplus beyond local needs. Typically, even short fallow-based rainfed agricultural systems have diverse tree components, for example, 167 different tree species are grown in two districts of western Kenya (Scherr 1995:794). However, the joint processes of market integration and agricultural intensification, especially mechanization, have often led to a significant reduction in the diversity of tree crops used and protected locally.

## CROP VARIETAL DIVERSITY

A second aspect of crop biodiversity is crop varietal diversity, for which the indicator selected is the proportion of cropped area grown with modern varieties (sometimes known as high yielding varieties or HYVs). The higher the adoption of modern varieties, the lower is the *spatial* diversity of varieties likely to be, but the higher is the *temporal* diversity likely to be, because of rapid turnover.

The rate and levels of adoption of modern varieties vary between crops, moisture regimes, and regions. For rice and wheat, at least, irrigation is synonymous with modern varieties *(see Table 24)*. Deepwater and upland rice, and arid and semiarid rainfed wheat rely heavily on traditional varieties, although the production shares involved are quite small. In more heterogeneous farming areas, such as hillsides, and where such complementary inputs as chemical fertilizer are not available, there is greater reliance on a diverse range of traditional varieties *(see Box 5)*.

In the 1990s, adoption of modern varieties of wheat, rice, and maize in developing countries had reached around 90, 70, and 60 percent respectively (Smale 2000; Morris and Heisey 1998). In Latin America, modern rice varieties (almost exclu-

sively irrigated varieties) leapt from 4 to 58 percent in around 20 years, in Asia from 12 to 67 percent, reaching in China (according to this data) a startling 100 percent (Byerlee 1996:698). However, there still may be a dominant use of landraces and traditional varieties in Sub-Saharan Africa for rice, and for maize in West Asia/North Africa, Asia (excluding China), Sub-Saharan Africa, and Latin America.

## CROP GENETIC RESOURCE CONSERVATION

Genetic improvement, by conventional breeding or biotechnology, depends upon access to germplasm that may be screened in the search for promising or desirable traits. The greater the diversity of the germplasm, the more likely the search will be successful. Some traits may be found in previously bred lines but often, particularly for such traits as pest and disease resistance, they are to be found in wild relatives *(see Box 6)*.

Modern approaches to conservation involve a combined *ex-situ* and *in-situ* strategy (FAO 1998). For many years, most official conservation has been *ex-situ*, primarily through genebanks set up by agricultural research institutions, universities, and other scientific agencies. The largest collections are in China, the United States, and Russia who hold some 300,000, 268,000, and 178,000 accessions, respectively, while the Consultative Group for International Agricultural Research (CGIAR 2000) collectively holds just over 10 percent of the world's *ex-situ* collection, with nearly 600,000 accessions (FAO 1998:98-99). Genebanks have several limitations. They tend to focus only on major crop species, and separate samples and germplasm from their natural ecosystem, preventing effective adaptation. Genebanks also require long-term commitment to ensure that seeds are periodically regenerated and that samples are properly characterized.

*In-situ* conservation programs decentralize seed selection and allow farmers to exchange and cross varieties, using community-based seed-maintenance programs for storing local and rare varieties. Farmers cultivate diverse varieties and landraces on their own farms and may collaborate with researchers in vari-

*Box 6*

## Biodiversity of *Phaseolus* Bean Species in Latin America

The most important grain legume for human nutrition is the bean. There are seven species (*Phaseolus vulgaris, P. lunatus, P. coccineus, P. polyanthus, P. purpuracens, P. glabella, and P. acutifolius*), which occupy 6 percent of the total agricultural area of Latin America. The greatest diversity of wild bean species is found in the arid zones of Mexico and northern Argentina and the humid, hillside areas of Central America and the Andean region. Map 20 shows the distribution of diversity of wild populations of common bean (*p. vulgaris*), the most commonly cultivated species, together with an agroclimatic characterization and an indication of the primary bean production areas.

However, the diversity of the cultivated species can also be analyzed at other levels, such as bean market classes from a consumer perspective, growth characteristics, and cultivar diversity within each production region. Genetic improvement of common bean in Latin America has been undertaken using breeding strategies designed with the following characteristics: (1) responds to consumer and market preferences for bean size, shape, and color and farmers' requirements for maturity and growth characteristics; and (2) overcomes constraints, mainly diseases. Excessive reliance by breeders on a few germplasm sources for disease resistance has led to a reduction in genetic diversity. Nevertheless, if we consider all the types and races of improved bean cultivars—traditional and wild varieties grown in these areas—the genetic diversity is higher than for most other crops.

Contrary to the situation in other regions with other crops, there has been no displacement of varieties by improved common bean cultivars in Latin America, therefore, there has been no biodiversity loss due to the introduction of new materials. On the contrary, because the genetic base for common bean cultivars is narrow at the intraracial level and the released materials contain new genetic combinations, which overcame some deficient cultivar traits, genetic variability has been broadened and made more useful.

The establishment of protected areas, urban growth, and other aspects of land use have important effects on agrobiodiversity. Map 21 shows how wild populations of common beans are distributed in relation to both protected and urban areas in the countries of Mesoamerica. Few protected areas contain significant areas of bean biodiversity, which could reflect the low emphasis given to agrobiodiversity as a planning criterion when siting protected areas. Likewise, many areas harboring agrobiodiversity are currently located in urban areas or in areas of easy access, thus exposing the germplasm to degradation or loss.

These spatial patterns are important when considering the appropriateness of *in-situ* or *ex-situ* conservation strategies. *In-situ* conservation is preferable for inaccessible areas with high levels of agrobiodiversity, while *ex-situ* conservation may be more suitable for areas rich in agrobiodiversity that are at high risk through exposure to urbanization and easy access.

*Sources*: Winograd, M. and Farrow, A. 1999. *Agroecosystem Assessment for Latin America.* Prepared for WRI as part of the PAGE project.

Voysest, O., M. Valencia, M. Amezquita. 1994. *Genetic Diversity Among Latin American Andean and Mesoamerican Common Beans Cultivars.* Crop Science. Vol. 34, 4:1100-1110.

etal selection for crop improvement. Storage facilities are in local communities, facilitating access. These techniques are particularly useful in countries where indigenous varieties are under threat (Thrupp 1998).

We have used the proportion of landraces and wild species in collections as an indicator of the extent to which crop, tree, and livestock biodiversity is being conserved. Table 25 summarizes the known holdings at the global level for the major food crops.

Looking to the future, both the WIEWS and ICIS databases support linkages to geoferenced data, so germplasm information of the type presented here will also become available in map form. Relevant germplasm indicators and monitoring approaches may be assessed under the auspices of the FAO International Undertaking for Plant Genetic Resources (FAO 2000b).

## AREA PLANTED TO TRANSGENIC CROPS

Using genetic engineering, scientists can directly insert specific traits into the genetic material of a crop or animal. Although farmers and scientists have been modifying crops and animals for centuries, genetic engineering allows much greater control over the modification process as well as new forms of genetic combination. In general, a single gene from an outside source (within or outside the species) containing coding for desired characteristics—such as herbicide resistance or an antibacterial compound—is inserted into the recipient organism (Persley and Lantin 2000). For example, frost resistance in tomatoes has been enhanced by using fish genes; the capacity to produce insecticidal proteins from *Bacillus thuringiensis* (Bt) has been developed for several major crops.

Biotechnology may bring about major advances, for example, raising the potential yield of crops by altering plant stomata

*Table 25*
## Genetic Diversity and Germplasm Collections for Major Food Crops

| Commodity | 1997 Harvested Area[a] (mha) | Landraces Number (x 1,000) | Landraces In Collections (percentage) | Wild Species Number (units) | Wild Species In Collections (percentage) | Ex-situ Major Collections (units) | Ex-situ Germplasm Accessions (x 1,000) | Ex-situ CGIAR Holdings (percentage) |
|---|---|---|---|---|---|---|---|---|
| **Cereals** | | | | | | | | |
| Bread Wheat | 210 | | 95 | 24 | 60 | 24 | 784 | 16 |
| Durum Wheat | 15 | 150 | 95 | 24 | 60 | 7 | 20 | 14 |
| Triticale | – | | 40 | | | 5 | 40 | 38 |
| Rice | 152 | 140 | 90 | 20 | 10 | 20 | 420 | 26 |
| Maize | 143 | 65 | 90 | | 15 | 22 | 277 | 5 |
| Sorghum | 44 | 45 | 80 | 20 | 0 | 19 | 169 | 21 |
| Millets | 37 | 30 | 80 | | 10 | 18 | 90 | 21 |
| Barley | 65 | 30 | | | | 16 | 484 | 5 |
| Oats | 16 | | | | | 20 | 222 | 0 |
| Rye | 11 | | | | | 8 | 287 | 0 |
| **Food Legumes** | | | | | | | | |
| Beans | 27 | | 50 | | 70 | 15 | 268 | 15 |
| Soybeans | 67 | 30 | 60 | | | 23 | 174 | 0 |
| Chickpeas | 11 | | | | 75 | 13 | 67 | 41 |
| Lentils | 3 | | | | 95 | 5 | 26 | 30 |
| Fava Beans | 3 | | | | 25 | 10 | 29 | 33 |
| Peas | 7 | | | | 0 | 18 | 72 | 0 |
| Groundnuts | 23 | 15 | | | 28 | 16 | 81 | 18 |
| Cowpeas | 7 | | | | 30 | 12 | 86 | 19 |
| Pigeon Peas | 4 | | | | 22 | 4 | 25 | 52 |
| Lupin | 1 | | | | | 10 | 28 | 0 |
| **Root Crops** | | | | | | | | |
| Potato | 19 | 30 | 95 | | 30 | 16 | 31 | 20 |
| Sweet Potato | 10 | 5 | 50 | | | 7 | 32 | 21 |
| Cassava | 16 | | 35 | | 29 | 5 | 28 | 30 |
| Yam | 3 | 3 | | | | 2 | 12 | 25 |
| **Other** | | | | | | | | |
| Sugar Cane | 19 | 20 | 70 | | | | | |

**Source:** Compiled from Evenson et al. 1998:2. (a) Crop harvested area: FAOSTAT 1999: Heisey et al. 1999.
**Note:** CGIAR: Consultative Group on International Agricultural Research.

and thus reengineering the photosynthesis process (Mann 1999). And industry is poised to introduce an array of "prescription foods", bioengineered to provide nutrients for those suffering from deficiencies, such as iron-enriched foods for people with anemia. Bioengineered plants could become "chemical factories", concentrating the production of particular enzymes or phyto-chemicals deemed to have positive effects on human health (DellaPenna 1999).

Environmental concerns about transgenic crops include the fear that genes introduced into crops will spread and become established in related native species, as happens with conventionally bred crops. In the case of genetically modified (GM) crops, the inserted genes are often derived from other taxonomic groups and give traits not previously present in wild plant populations. The concerns expressed, therefore, are about the un-

known consequences of accidentally spreading these new genes into wild populations.

Genetically modified, herbicide-tolerant crops raise other concerns. For example, the broad-spectrum herbicides applied to them may be far more damaging to farmland ecosystems than the selective herbicides they might replace. Using these herbicides in the growing season may also increase spray drift onto adjacent habitats, such as hedgerows and watercourses. Furthermore, it is feared that the agricultural intensification made possible by GM crops will further threaten declining wildlife populations (Johnson 2000), and that insects may develop resistance to Bt crops, increasing the threat to other crops (Gould and Cohen 2000).

On the other hand, GM technology could be directed to biodiversity-enhancing objectives. Examples might include achieving insect resistance by altering the physical characteristics of plants (e.g., increasing hairiness) to reduce insecticide use, or developing crops that can tolerate high levels of natural herbivory, yet remain viable (Johnson 2000).

Concerns about the environmental and food safety of genetically modified organisms (GMOs) suggest that the area planted with genetically modified crops is a useful indicator to monitor. The global area planted with transgenic crops, some 82 percent of which was in OECD countries, increased from only 1.7 million hectares in 1996 to 39.9 million in 1999. The seven principal transgenic crops grown in 1998 were (in descending order of area) soybean, maize, cotton, canola (rapeseed), potato, squash, and papaya (Persley 2000:26 quoting James 1999).

## Enhancing the Capacity of Agroecosystems to Support Biodiversity

A fundamental issue addressed in the *Status of Land Conversion to Agricultural Use* section was the loss of natural habitat through land conversion to agriculture. Thus, strategies that minimize the demand for agricultural land are potentially beneficial for biodiversity. In this regard, one important strategy is to continue improving agricultural productivity. A recent study by Goklany (2000:160-161) calculated that yield increases since 1961 may have forestalled the conversion of an additional 3.3 billion hectares of habitat globally to agricultural uses (including 1.0 billion hectares of cropland). He suggests that if productivity does not increase over the next 50 years, cropland would have to increase by 1.7 billion hectares to feed the predicted population of 9.6 billion people by the year 2050. But, an increase in productivity of 1 percent per year would reduce the required crop area expansion to less than 400 million hectares (Goklany 2000: 183). A second study by Nelson and Maredia (1999:29) estimates more conservatively that 170–420 million fewer hectares of tropical land were converted to agriculture because of productivity-increasing research carried out by the CGIAR on the ten major food commodities between the 1960s and 1990s. There may be offsetting losses, however, if the new ways of enhancing productivity are more damaging to biodiversity.

A related strategy being promoted in many high-income, food-surplus countries is to encourage land conversion back to natural habitat. Retirement of marginal agricultural lands has been achieved through a combination of land use controls and subsidies for biodiversity conservation (e.g., the Conservation Reserve Program in the United States, and the Set-aside program of the European Union). The economic value of natural biodiversity may justify reconversion, for example, freshwater fisheries in the Mekong Basin (ICLARM 1999) or water quality in urban watersheds. Various transfer payment schemes are also being devised to promote such objectives (McNeely and Scherr forthcoming).

Both the above strategies are conditional on the agricultural productivity of land (yield per hectare), an indicator used in assessing the condition of agroecosystems to provide food *(see Food, Feed, and Fiber section)*. That indicator could be further developed in this context. First, by including both cropland and pasture in the land productivity measure, since clearing for pasture is the dominant cause of agricultural land conversion. Second, by constructing a new indicator as a ratio of the growth rate of land productivity to the growth rate of population within the same region. Ratios less than one would imply various degrees of land conversion pressure from a food production perspective; ratios greater than one would imply various degrees of potentially surplus agricultural land.

In many land-scarce countries, and where agricultural areas are located in or near important centers of biodiversity, strategies to enhance biodiversity must address the design of agricultural production systems and the configuration of agricultural landscapes (McNeely and Scherr forthcoming).

With regard to enhancing the quality of natural habitats in farming areas, research is underway to identify minimum patch size and other resource requirements to sustain particular wild species or species clusters (for example, Ricketts et al. 2000; Wilson 1992, the Forest Fragments Project). Agroforests, improved fallows, and other systems are being developed to "mimic" the structure and composition of native habitats in the tropics (Leakey 1999; Lefroy et al. 1999). Strategic use of trees, protected corridors, hedgerow, and wetlands, among others is becoming part of the design of landscapes that serve both production and environmental functions (Forman 1995). Farmers may be encouraged to adopt these strategies as a result of regulation, perceived farming benefit (such as protecting pollinators), personal involvement in community planning initiatives, tax incentives, or direct payments for habitat protection (McNeely and Scherr forthcoming). A notable case of technology influencing the nature of landscapes is evident in the requirements for direct payment in the European Union set-aside scheme, where plots must be no less than 20m wide to ensure they can be detected by satellite monitoring (MAFF 2000).

There is mounting evidence that farmers increasingly protect or establish trees on farms for economic reasons: to supply subsistence needs for fuelwood, building materials, fencing, and other products; or to market for cash income. This situation reflects the global revolution in forest product supply, in which domesticated tree production is replacing the shrinking natural forests that were formerly the major sources of those products, and local markets of selected products are evolving to regional or national markets (Dewees and Scherr 1996). Ongoing research and market development initiatives promise to increase greatly the rate of domestication of tree species (Leakey 1999) *(see Box 7)*.

*Box 7*

## Tree Diversity on Farms

In recent years, a new movement has arisen to domesticate some wild trees from tropical forests and woodlands for their fruits, timber, and other products. This initiative has been hailed as the start of a "Woody Plant Revolution" to follow the Green Revolution. The trees targeted for domestication are species whose products humans have traditionally collected, gathered, and used from the wild. These products are important to many people around the tropics for food and nutritional security and are traded in long-established or emerging markets but science has largely overlooked these "Cinderella" species.

The miombo woodlands of southern Africa, for example, are a reservoir of biodiversity—over 50 fruit tree species bear widely used edible fruits. Many serve as food reserves and enhance food security in times of famine. Some contribute to cash income of farmers, with important regional markets for jams, juices, and alcoholic beverages. With wide-scale deforestation, many of these wild fruit species are threatened with extinction. Farmers and researchers are enthusiastic about the potential for domesticating some of these species for farm production. However, to make economic production feasible, efficient methods must be found for propagation, and traits such as tree size, early fruiting, reduction of thorns, and resistance to pests and disease, must be improved.

Increasing the quality, number, and diversity of domesticated trees to fill the niches in farmers' fields can not only benefit farmers and preserve specific wild tree species, but also protect other wild biodiversity. Research now underway in the tropics to integrate tree plots strategically in agricultural landscapes, and to improve and develop new agroforest systems (polycultures of economically valuable plants), offers promising opportunities to provide habitat for many more wild species than do conventional agricultural fields. There is a perceived contradiction between making genetic gains through the selection of superior trees (so as to make tree production economic and attractive to farmers) and the need to maintain genetic diversity. In reality, these two situations can be compatible, providing they are both part of a wide risk-averse strategy for domestication.

Many efforts are now underway to preserve biodiversity by making wild tree species more attractive for farmers to grow, through technical research, farmer extension, and market development. Some are already beginning to pay off.

- *Ziziphus Mauritania*, used for making fences, was threatened by overexploitation in drylands and had disappeared from much of the northern fringe of its range in the Sahel. Researchers developed improved cultivars of ziziphus, focusing on the consumption and market potential of its vitamin-rich fruit. Ziziphus is now grown in farm hedges for multiple functions.

- Farmers in Cameroon were interested in cultivating the popular bush mango (*Irvingia gabonensis*), because strong markets already existed for the fruits, kernels, and bark from the wild trees. By developing a new propagation method, known as marcotting, the time required for the bush mango to fruit has been reduced from 10-15 years to 3-5 years.

- Western science "discovered" in the mid-1960s that the bark of the *Prunus africana* was an effective treatment for prostate disorders. Burgeoning demand threatened the slow growing species—grown only in the moist highlands of Africa—with extinction. New techniques of molecular analysis have accelerated progress in selecting for improved traits (e.g., more reliable seeding, seed viability) from wild genetic diversity, and in grafting to reduce time to maturity in seed orchards.

- Capirona (*Calycophyllum spruceanum*) is a fast-growing Peruvian forest tree that produces valued hardwood for timber and poles, as well as firewood. Although farmers are interested in growing the tree on their farms, capirona's seed had never been systematically collected to evaluate genetic differences in tree growth, wood quality, and other commercial characteristics. On-going research since 1996 has already identified provenances with double the standard growth rate of capirona trees.

**Sources:** ICRAF 1999. *Paths to Prosperity*. International Center for Research in Agroforestry, Nairobi, Kenya; Leakey, R.R.B. 1999. Win: Win Land Use Strategies for Africa: Matching Economic Development with Environmental Benefits Through Tree Crops. Paper presented at USAID Sustainable Tree Crop Development Workshop, "Strengthening Africa's Competitive Position in Global Markets," Washington, D.C. 19-21, October; Leakey, R.R.B. and Newton, A.C. 1994. Domestication of Tropical Trees for Timber and Non-timber Forest Products, MAB Digest No. 17, UNESCO, Paris.

The prospects for increasing the number of wild species within agroecosystems seem promising, although progress so far is limited. In addition to the opportunities related to landscape structure suggested above, technological change could potentially reduce the damage currently caused by agrochemicals and poor tillage systems, as evidenced by the dramatic expansion worldwide in the use of integrated pest management, reduced tillage systems, and organic farming. Policy action is supporting such trends. Examples are regulations restricting pesticide, fertilizer, and herbicide application and cutting on set-aside areas in the European Union to promote colonization by wild flora and fauna (for example MAFF 2000).

It is difficult to predict the broad prospects for enhancing agricultural species biodiversity. On the one hand are unprecedented international initiatives to conserve germplasm, to domesticate tree species, and to develop economically viable technologies for polyculture systems. These initiatives suggest the development of more diverse production systems where the currently dominant cereals play a lesser role. On the other hand, many economic, technical, marketing, and policy factors continue to reward specialization in production.

## Summary of Indicators and Data

This assessment leaves many difficult questions unaddressed. Agriculture is, by definition, an exercise in biological specialization, so improving agriculture's poor record with regard to biodiversity loss is a complex challenge. There is a need to produce food as cost-effectively as possible, but also to ensure that biodiversity loss is reflected as an element of the true costs of production. Biodiversity loss, particularly in commercial agriculture, is not widely recognized as important and thus not perceived to be in the private interests of farmers to remedy. However, many biodiversity losses in soil microorganisms, pollinators, carnivorous insects, and other species, are counter productive to enhancing agricultural productivity, that is, it would pay well-informed farmers to avoid such losses.

Without a clearer understanding of the dependencies between flora, fauna, and human action within and beyond agriculture, it is very difficult to develop a more strategic approach to tackling biodiversity issues in farmland. Food needs, and thus agricultural development, will continue to involve trade-offs with one or another aspect of biodiversity. The significant challenge to agriculturalists and ecologists is to avoid such trade-offs where possible and to mitigate impacts where not. This challenge will likely need continued mediation through conservation policy. From the perspective of environmental and agricultural policy design, it is important that we gain a better understanding of what is at stake and, hence, are better placed to design biodiversity interventions compatible with specific social concerns, including long-term food security.

# CARBON SERVICES

Carbon, an essential component of all life forms, is constantly being absorbed, released, and recycled by a range of natural and human-induced biological and chemical processes. Agriculture both influences and is affected by this global cycle. Of particular importance is the process of photosynthesis in which plants absorb atmospheric carbon as they grow and convert it into biomass. Furthermore, when plant residues and roots decompose, the carbon they contain is transformed primarily into soil organic matter and carbon based gases. One product, soil organic matter, is particularly critical in conditioning soil quality and, hence, the capacity of agroecosystems to provide a number of agricultural and environmental goods and services.

The cumulative impact of human activities, including agriculture, has been to significantly increase the atmospheric concentration of the so-called greenhouse gases, such as carbon dioxide ($CO_2$) and methane ($CH_4$). Higher concentrations of greenhouse gases have induced (or accelerated) the process of global warming—the gradual increase in the earth's surface temperature that is also having an observable impact on the amount, variability, and spatial pattern of precipitation and, through such changes, on agricultural production.

Thus, from an agroecosystem perspective there are two related groups of carbon goods and services. The first is the direct, on-site contribution of carbon to the production of plant (and hence, animal) biomass and improved soil fertility. Farmers and researchers seek farming systems that increase crop and pasture productivity, reduce soil degradation, and improve soil organic matter in mutually reinforcing ways. These soil productivity aspects of carbon services are dealt with in the section on Soil Resource Conditions. This section focuses on the second group of carbon services linking agricultural production and global warming.

This section reviews: the contribution of agroecosystems to terrestrial carbon storage, the status and trends of carbon emissions from agroecosystems, and the potential for enhancing the capacity of agroecosystems to provide global carbon services. Such services include increasing carbon sequestration, reducing carbon emissions, and increasing long-term carbon storage.

## Status of Carbon Stocks

Through the 1992 U.N. Framework Convention on Climate Change (FCCC) and subsequent agreements, primarily the 1997 Kyoto Protocol, the international community has resolved to take action to reduce future anthropogenic (human-induced) contributions to greenhouse gases (UN 2000). Beyond reducing emissions from fossil fuel combustion, a further option for mitigating the build-up of greenhouse gases is to adapt land use practices

*Table 26*

## Vegetation and Soil Organic Carbon Storage within the PAGE Agricultural Extent

| Agroclimatic Zone[c] | PAGE Agricultural Extent[a] | | Total Carbon Storage[b] | | | Carbon Storage Density | | |
|---|---|---|---|---|---|---|---|---|
| | | | Vegetation (Low-High) | Soils (Mean) | Total[d] (Low-High) | Vegetation (Low-High) | Soils (Mean) | Total[d] (Low-High) |
| | *(m sq km)* | *(% total)* | *(GtC: gigatons of carbon)* | | | *(Mt C/ha: metric tons per hectare)* | | |
| Temperate | | | | | | | | |
| Humid/Subhumid | 6.4 | *17.8* | 12-27 | 81 | 93-109 | 18-42 | 127 | 145-169 |
| Semiarid/Arid | 7.0 | *19.3* | 11-28 | 78 | 90-106 | 16-40 | 112 | 129-152 |
| Moderate Cool/Cool/Cold Tropics | | | | | | | | |
| Humid/Subhumid | 0.9 | *2.5* | 2-6 | 9 | 11-15 | 20-71 | 99 | 119-170 |
| Semiarid/Arid | 0.5 | *1.4* | 1-3 | 4 | 5-7 | 17-55 | 76 | 93-131 |
| Moderate Cool/Cool/Cold Subtropics | | | | | | | | |
| Humid/Subhumid | 4.5 | *12.4* | 9-21 | 48 | 57-69 | 20-47 | 107 | 127-154 |
| Semiarid/Arid | 2.5 | *6.8* | 3-8 | 18 | 21-26 | 11-31 | 74 | 84-105 |
| Warm Subtropics and Tropics | | | | | | | | |
| Humid/Subhumid | 8.6 | *23.8* | 17-59 | 83 | 100-142 | 20-68 | 96 | 116-165 |
| Semiarid/Arid | 5.6 | *15.4* | 7-19 | 42 | 49-60 | 12-33 | 75 | 87-109 |
| Boreal | 0.3 | *0.8* | 1-3 | 5 | 6-8 | 41-101 | 181 | 222-283 |
| Total PAGE Agricultural Extent | 36.2 | | 62-173 | 368 | 431-542 | 17-47 | 102 | 119-148 |
| Global Total[f] | | | 268-901 | 1,555 | 1,823-2,456 | | | |
| Agricultural Percentage of Total | | | *23-19* | *24* | *24-22* | | | |

**Source:** IFPRI calculation based on: (a) GLCCD 1998 and USGS EDC 1999, (b) WRI estimate of vegetation carbon stocks based on Olson et al. 1983 *(see note)*, and IFPRI estimate of soil carbon stocks based on FAO 1995, Batjes 1996 and Batjes 2000 *(see note, Table 19)* and (c) FAO/IIASA 1999.
**Notes:** The map of carbon stored in above- and below-ground live vegetation (1km by 1km resolution) is based on estimates developed by Olson et al. (1983) for the dominant vegetation types found in the world's major ecosystems and applied to the Global Land Cover Characteristics Data base (GLCCD). This method applied the average vegetation carbon storage value of a mosaic to each element of the mosaic. This approach overestimates, often significantly, the amount of carbon in the agricultural component of the mosaic. For example, an average carbon density for an area containing 50 percent broadleaf forest (120 t/ha carbon) and 50 percent rice paddy (30 t/ha carbon) would be 75 t/ha. In our estimates, we applied just the agricultural carbon density (in this case 30 t/ha) when computing the carbon storage contribution of the agricultural share of the mosaic. The soil carbon storage values relate to the upper 100 centimeters of the soil profile. For further details on the calculation of carbon stored in soils, see note, Table 19. (d) The total low estimate is the low estimate for vegetation plus the mean soil estimate; the total high estimate is the high estimate for vegetation plus the mean soil estimate. (f) Carbon storage values for vegetation include estimated carbon stores for Greenland and Antarctica. Soil carbon storage data for Greenland and Antarctica were largely incomplete and were excluded.

so as to increase the amount of carbon in terrestrial ecosystems. Thus, the amounts of carbon stored in agroecosystem vegetation and soils are important indicators of agriculture's contribution toward limiting climate change.

The global carbon stocks of most relevance to agriculture are the roughly 750–800 GtC (gigatons or thousand million tons of carbon) contained in Earth's atmosphere and the 2,000-2,500 GtC held in organic forms in terrestrial systems. Terrestrial organic carbon stocks can be separated further into that part stored in vegetation and the much larger amount stored in soils (IPCC 1996a; IPCC 2000:30).

In order to generate compatible estimates of carbon storage, PAGE researchers made a joint assessment of carbon stocks for the three terrestrial ecosystems included in the PAGE study:

forests, grasslands, and agriculture.[22] A key reason for harmonization was to avoid double counting or omission of areas by the PAGE ecosystem teams, particularly where complex land use patterns (the so-called land cover "mosaics" that could include, say, forest and cropland) were known to exist.

### VEGETATION CARBON STORAGE

The PAGE assessment linked two data sources: the global assessment made by Olson et al. (1983) of the above- and below-ground carbon storage in live vegetation; and the 1 km. resolution Global Land Cover Characteristics Database (GLCCD 1998) interpreted to delineate global ecosystems (USGS EDC 1999b). Olson and his colleagues estimated carbon storage densities for the dominant vegetation types found in the world's major ecosystems, and applied these densities to a contemporary vegetation map of the world. For the PAGE assessment, researchers at WRI linked each ecosystem complex in the GLCCD to the high and low carbon densities used for equivalent vegetation types in the Olson study.[23]

Olson's vegetation schema is extremely limited with regard to agriculture, a limitation offset in part by the much larger differences in carbon density between, rather than within, major vegetation classes. For example, the average carbon density for tropical/subtropical broad leaved humid forest used by Olson et al. (1983) was 120 t ha$^{-1}$ (ranging from 40 to 250 t ha$^{-1}$), while for paddy rice and tropical (natural) savanna pastures the corresponding figures were 30 t ha$^{-1}$ average (range 20-40 t ha$^{-1}$ for paddy and 20-50 t ha$^{-1}$ for savanna). More recent assessments of agricultural vegetation suggest croplands can contain 20-60 t ha$^{-1}$ carbon and tropical agroforestry systems 60-125 t ha$^{-1}$ (Dixon et al. 1993:164).

Because of the disparity between carbon storage in agricultural and many other vegetation types, the WRI approach to estimating carbon storage in vegetation was extended in the case of land cover mosaics containing agriculture. For a mosaic containing, say, 50 percent broadleaf forest (120 t/ha carbon) and 50 percent rice paddy (30 t/ha carbon) the average vegetation carbon density is 75 t/ha. Instead of applying the mosaic average, and thus overstating the carbon storage of agricultural vegetation, we used only the agricultural carbon density (in this case 30 t/ha) when computing the carbon storage contribution of agricultural vegetation in mosaics. Map 22 displays the global variation in the density of carbon storage in vegetation within the PAGE global extent of agriculture, based on Olson's high carbon densities for agricultural land cover types.

The vegetation carbon storage estimates presented here should, therefore, be treated as indicative only. The average density across the PAGE agricultural extent is 17 metric tons per ha (low) to 47 metric tons per ha (high) and the corresponding total carbon storage in agricultural vegetation lies within the range of 62 to 173 gigatons. Table 26 shows a breakdown of total carbon and average carbon density in vegetation by major agroecosystem groups, revealing generally higher vegetation densities in more humid and warmer environments. The total stock of vegetation carbon is greater in the humid and subhumid, warm subtropics and tropics, and in temperate agroecosystems.

In practice, the value of agroecosystem vegetation for carbon storage is limited. Not only are agricultural biomass densities lower than those of forests and natural grasslands, but biomass is regularly harvested and used in ways that release the stored carbon. Only if the share of deep-rooted, woody, or tree crops was significantly increased in agricultural lands would agriculture notably contribute to global vegetation carbon storage over the longer-term.

## SOIL ORGANIC CARBON (SOC) STORAGE

Soil microorganisms decompose dead roots and above-ground residues of plants and animals. Decomposition results in the release of carbon as $CO_2$ or $CH_4$, but also in the formation of soil humus or organic matter, the principal store of organic carbon in soils. Batjes (1996) interpreted 4,353 soil profiles contained in the World Inventory of Soil Emission Potentials (WISE) database compiled by the International Soil Reference and Information Centre (Batjes and Bridges 1994). He linked these profiles to the digital FAO-UNESCO Soil Map of the World (FAO 1991) to derive estimates of the global variation of organic carbon stored at depth intervals of 30cm, 100cm and 200cm. The estimates were made by summing the SOC content of the soil types found in each 0.5 by 0.5 degree grid cell based on a weighting of SOC values by soil type area shares. The analysis provided two values for each depth: assuming the median stone content of soil and assuming a "stone free" soil, which yields a higher soil carbon figure. Batjes (1996:158) calculated the global stock of organic carbon in the first 100cm of the soil layer, the depth range most directly involved in interactions with the atmosphere and most sensitive to land use and environmental change, to be 1,462-1,548 GtC. This figure is compatible with a separate estimate of 1,576 GtC made by Eswaran et al. (1993:193).

For the purposes of the PAGE study we repeated Batjes analysis using a more recent 5 x 5 minute grid of the Soil Map of the World[24] (FAO 1995) together with average SOC content values taken from Batjes (1996 and 2000). We did not adjust for stoniness, so the PAGE estimate of global organic carbon of 1,555 Gt in the upper 100 cm of soil corresponds to Batjes' estimate of 1,548 Gt.

Map 23 presents the PAGE estimate of organic carbon storage in the top 100cm of the soil layer within the agricultural extent. The map clearly shows the greater amounts of organic carbon storage at higher latitudes. Higher rainfall and higher temperatures, characteristic of much of the tropics, tend to wash away, leach, and more rapidly recycle organic matter. The map also shows areas such as in Malaysia, Sumatra, Borneo, and Sulawesi where highly organic tropical soils (peats) occur. The importance of organic matter in soil fertility is witnessed by the strong spatial correspondence between the world's most productive agricultural lands and areas with higher organic carbon content. Table 26 summarizes the spatial variation of soil carbon estimates for major agroecosystem groups. Global average soil organic carbon density is estimated at 102 metric tons of carbon per hectare of land within the PAGE agricultural extent, considerably more than even the high estimate for vegetation. The higher carbon densities in the soils of boreal and temperate zones, and in more humid areas across all zones are also apparent. The total global store of soil organic carbon within

*Figure 14*

## Net Carbon Dioxide Emissions from Land Use Change: 1850-1998

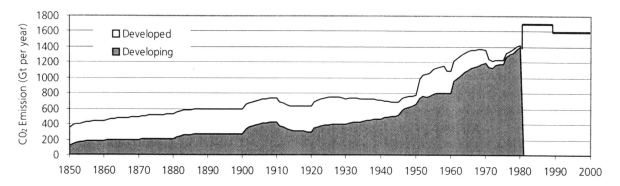

**Source and Notes:** Houghton and Hackler 1995 (time series from 1850-1980). 1980-89 average of 1,700 GtC per year and 1989-98 average of 1,600 GtC per year from IPCC 2000. Developing includes China, South and Central America, North Africa and Middle East, tropical Africa, South and Southeast Asia. Developed includes North America, Europe, the former Soviet Union, and Pacific developed region. Houghton and Hackler estimated (1) the carbon flux in the atmosphere from clearing or degradation of vegetation, cultivation of soils, decay of dead vegetation, and (2) the recovery of abandoned lands.

the PAGE agricultural extent is estimated at 368 GtC with 43 percent located in temperate agroclimatic zones.

Map 24 shows a newer, more detailed, estimation of soil carbon content for South America. The soil boundaries and descriptions for this map are taken from the Soil and Terrain (SOTER) database of Latin America (FAO 1999c) linked to the WISE soil profile database (Batjes and Bridges 1994). The enhanced interpretation potential of this new generation of data and analysis is revealed by comparing the insets for the Pampas, northern Argentina, Uruguay, and the southern-most part of Brazil. The bottom inset is taken from the SOTER-derived map to the left, while the top inset represents the equivalent areas from the PAGE estimate of soil organic carbon. Differences arise from changes in soil organic carbon levels along with improvements in the underlying soil information as a consequence of considerable national and international investment in soil data collection, interpretation, and mapping.

## Status and Trends of Greenhouse Gas Emissions

From 1989-98, human activities contributed an estimated average of 7.9 GtC per year of $CO_2$, the primary greenhouse gas, to the atmosphere. Much of that emission was reabsorbed into oceans (2.3 GtC) and terrestrial systems (2.3 GtC), leaving a net global increment into the atmosphere of about 3.3 GtC per year. Analysts believe that land use changes and practices, predominantly in the tropics, contributed some 1.6 GtC (20 percent) of the $CO_2$ emissions (IPCC 2000:32). Agriculture-related land use activities that emit $CO_2$ include the following: the clear-

ing of forest and woody savanna for agriculture; the deliberate burning of crop stubble and pastures to control pest and diseases and promote soil fertility, drainage, cultivation, and soil degradation (Lal 2000; Rosenzweig and Hillel 2000). Figure 14 shows the dramatic increase in land use related $CO_2$ emissions since the 1850s and the extent to which developing country emissions now dominate. In many developed countries, reforestation and related land management policies appear to be offsetting other forms of agricultural $CO_2$ emission. Houghton et al. (1999:575-577) show, for example, that since 1945 the United States has been a net carbon sink in terms of land use change effects—primarily through reforestation of agricultural land, fire suppression, and agricultural set-aside programs—to an extent that may have offset between 10 and 30 percent of U.S. fossil fuel emissions. The cumulative effect of anthropogenic emissions is believed to have raised the atmospheric concentration of $CO_2$ from its preindustrial (mid-1800s) level of about 285 to about 366 ppm in 1998, increasing at a more rapid rate with the passing of each decade (IPCC 2000:29).

Table 27 summarizes the major agricultural sources of carbon-based greenhouse gases. A rough estimation suggests that agriculture may be contributing around 20 percent of the greenhouse gas effects of global warming (including $N_2O$), with methane emissions becoming the more significant source of carbon-based gases (IPCC 1996b:748). In turn, analysts predict that carbon dioxide emissions from agriculture-related sources will decrease between 1990 and 2020 (Sombroek and Gommes 1996). Although methane, $CH_4$, is much less prevalent than $CO_2$, it is around 20 times stronger per molecule in its global warming potential (IPCC 1996a). Methane concentrations in the at-

*Table 27*

## Greenhouse Gas Emissions Related to Agriculture

| Agriculture-related Activity[a] | Carbon-based Greenhouse Gas Emitted | | | |
|---|---|---|---|---|
| | Methane[b] | | Carbon Dioxide[c] | |
| | *(MtC/year)* | *(percent)* | *(MtC/year)* | *(percent)* |
| Land Use Change and Cultivation Practices: | | | | |
| Biomass Burning | 22 (11-33) | 6 | 1,600 ± 800 | 20 |
| Rice Paddies | 50 (20-60) | 13 | | |
| Livestock | 94 (75-118) | 25 | | |
| Enteric Fermentation (Ruminants) | 80 (65-100) | 21 | | |
| Waste Management and Disposal | 14 (10-18) | 4 | | |
| Total Agriculture-related Emissions | 166 (106-201) | 44 | 1,600 ± 800 | 20 |
| Total Anthropogenic Emissions | 375 | 100 | 7,900 ± 800 | 100 |
| | | | | |
| Cumulative Emission | ~24 GtC[d] | | 136 ± 55 GtC[e] | |
| Annual Rate of Increase, 1995[f] | ~0.9 % per year | | ~0.5 % per year | |
| Direct Global Warming Potential (GWP)[g] | 21 | | 1 | |

**Sources and Notes:** (a) All land use change and biomass burning are conservatively attributed to agriculture-related activities. (b) Methane emission estimates from Table 23-11 IPCC 1996b. (c) IPCC 2000. (d) Cumulative emission of all anthropogenic methane from 1860 to 1994 (calculated from Stern and Kauffmann 1998). (e) Cumulative emission from land use change and cultivation -1850 to 1998 (IPCC 2000). (f) Houghton 1997. (g) Global warming potential (GWP) is the effectiveness of a greenhouse gas to trap heat in the atmosphere, relative to $CO_2$, in this case over a 100-year time frame.

mosphere have increased from preindustrial levels of 700 ppb to 1721 ppb in 1994. Concentrations are currently increasing at around 4-8 ppb per year (IPCC 2000:33). Furthermore, agriculture is the largest anthropogenic source of methane, now contributing around 44 percent of such emissions through livestock production, ruminant digestion by-products, and animal waste management and disposal (~25 percent), paddy rice cultivation (~13 percent), and biomass burning (~6 percent) (IPCC 1996b:764). Figure 15 traces the growth of human induced methane emissions and shows how livestock has become the largest agriculture-related source. Although methane emission growth rates have been declining in recent years, the on-going surge in demand for livestock products could reverse that trend (Delgado et al. 1999).

Table 28 indicates that ruminant livestock growth rates since 1980 are over twice that of rice areas, and that Asia and Sub-Saharan Africa are experiencing very high and sustained livestock growth rates. Forest area statistics are probably less reliable (and FAO ceased to report them in 1994) but suggest a slowing in deforestation at the global level and indicate that net reforestation is occurring in 5 of the 10 regions.

## Enhancing the Capacity of Agroecosystems to Provide Carbon Services

In recent times, some high-income, food-surplus countries have retired land from agricultural production and reforested previ-

ously cleared areas. However, future global food needs suggest these might not be feasible long-term approaches unless technological advances, as yet unknown, can significantly improve the productivity of existing agricultural land. Low-income and land-scarce countries will seldom have the option of taking land out of agriculture. Thus, the focus should be on change within the agricultural sector that might enhance the capacity to reduce or prevent carbon emissions or increase carbon sequestration and storage. Because farmers will play a role in implementing change, the most viable technical approaches to improving carbon services will be those that simultaneously improve agricultural productivity or provide other added value from a farmer perspective. Fortunately, most agricultural practices that help stabilize or reduce the concentration of greenhouse gases are beneficial to agriculture, for example, enhancing soil organic matter that improves soil fertility and structure, and reduces soil erosion.

Following land conversion into agriculture, cultivation practices generally accelerate the decomposition of organic matter, particularly at higher levels of temperature and humidity, although carbon losses in cropland are generally greater than those in pastures (IPCC 1996b). Practices that involve removing crop residues and insufficient nutrient replenishment or that lead to erosion, overgrazing, compaction, burning, acidification, or salinization also serve to accelerate the loss or decomposition of organic matter (Lal 2000). Analysts estimate annual losses of soil carbon from agricultural soils to be around 500 MtC (Lal 2000:64) for tropical soils, and around 800 MtC globally (Schlesinger 1990). Historically, the accumulated loss of car-

*Figure 15*

## Global Anthropogenic Methane Emissions by Source: 1860-1994

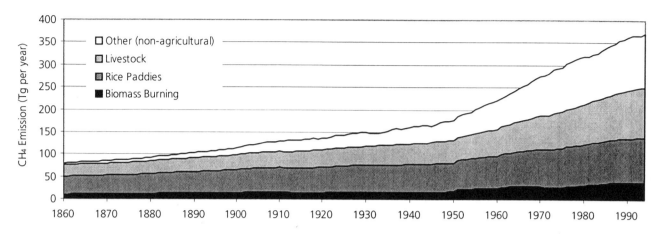

**Source:** Stern and Kaufman 1998.

bon from agricultural soils has been estimated as some 45-55 GtC—equivalent to 6-7 percent of the atmospheric carbon stock. Some researchers have suggested that this is the theoretical limit of organic matter that could be restored to those soils (Rosenzweig and Hillel 2000:50; IPPC 1996b:751).

Any cropping system strategies that promote increased production and retention of surface litter, involve less cultivation and burning, contain deep-rooted crops or increased tree components, and conserve soil and moisture would enhance the soil's organic carbon content. Controlling soil erosion in the United States alone could reduce $CO_2$ emissions by an estimated 12-22 GtC per year (Lal et al. 1998). Minimum or no tillage cultivation practices embody many such desirable qualities. Farmers are attracted to minimum tillage cultivation because it reduces production costs and improves soil and water conditions. It has an added benefit of reducing fossil fuel emissions; one study estimates this reduction to be about 45 percent compared to conventional tillage practices (Frye 1984). Minimum tillage appears to be beneficial under temperate to tropical conditions and for a broad range of annual crops. It has been shown to augment soil organic carbon by up to 50 percent (Bayer et al. 2000; Ismail et al. 1994; Kern and Johnson 1993:208).

Other strategies to reduce carbon emissions, particularly with regard to methane emissions, include better management of rice paddies and livestock waste, and improved feed quality for ruminant livestock to reduce the volume of methane produced during digestion. Improved pastures can also generate considerable soil carbon benefits. Fisher et al. (1994:237) showed how deep-rooted grass and legume forage crops used in the South American savannas accumulated up to 270 t ha[-1] of carbon in

the top 100 cm of soil, some 36 percent higher than under natural pasture conditions. They estimated that adopting such technologies had increased soil carbon sequestration by an estimated 100-507 MtC per year in the region (Fisher et al. 1994:236). Furthermore, much of this carbon was stored deeper in the soil profile below the plow layer.

Increased use of tree and woody biomass crops can also improve carbon storage in both soil and vegetation. Another noteworthy strategy is that of producing biofuel crops that recycle atmospheric carbon, offsetting the need to release an equivalent amount of new $CO_2$ into the atmosphere by burning fossil fuels (IPCC 1996b). Because biofuel production would likely compete with conventional agricultural land uses for food production, the strategy appears best suited to land- and food-surplus countries or regions or on lands unsuitable for sustainable annual crop production.

New incentives have emerged from the Kyoto Protocol that allow "certified emission reductions" from land-use related changes to be credited to the emission reduction commitments of signatory countries (a development that has already initiated speculative concepts of carbon "farming"). This process includes a mechanism by which industrialized countries can invest in carbon emission reductions in developing countries to gain credits. Typically, such joint initiatives might involve preserving or planting forest or other land cover with a high carbon-storage capacity. Because some agroforestry systems have high carbon-sequestration capacity, they are also candidates for such initiatives, a move that would further enhance the economic value of agroforestry. Furthermore, while increasing the carbon content of tropical agricultural soils appears to hold promise, some progress is already being made in the case of reducing methane

*Table 28*
## Growth in Agriculture-Related Drivers of Carbon Emissions, 1961-98[a]

| Region | Annual Growth Rates | | | | | |
| --- | --- | --- | --- | --- | --- | --- |
| | Rice Area[b] | | Ruminant Livestock[c] | | Forested Area[d] | |
| | 1961–79 | 1980–98 | 1961–79 | 1980–98 | 1961–79 | 1980–94 |
| | (percentage per year) | | | | | |
| North America | 2.84 | 0.35 | 0.06 | -0.74 | -0.13 | 0.26 |
| Latin America and the Caribbean | 2.81 | -1.60 | 1.32 | 0.70 | -0.38 | 0.09 |
| Europe | 1.30 | 0.95 | 0.04 | -0.28 | 0.52 | 0.09 |
| Former Soviet Union | 9.53 | -1.81 | 1.02 | -1.99 | 0.00 | -0.83 |
| West Asia/North Africa | 1.60 | 2.40 | 0.76 | -2.13 | 0.19 | 0.00 |
| Sub-Saharan Africa | 2.83 | 2.74 | 1.95 | 1.96 | -0.07 | -0.10 |
| East Asia | 1.34 | -0.49 | 1.35 | 3.44 | -0.57 | 0.40 |
| South Asia | 0.85 | 0.43 | 1.15 | 1.61 | 0.76 | 0.21 |
| Southeast Asia | 0.76 | 0.97 | 0.68 | 2.30 | -0.56 | -0.26 |
| Oceania | 7.85 | 1.12 | 0.07 | -0.98 | 0.00 | -0.02 |
| World | 1.16 | 0.33 | 0.96 | 0.76 | -0.15 | -0.11 |

**Source:** IFPRI calculation based on FAOSTAT 1999.
**Notes:** (a) Forest area to 1994 only. (b) Includes both irrigated and rainfed rice. (c) Relates to enteric fermentation and ruminant wastes–waste from other livestock is excluded. (d) Forest statistics reflect net area changes but clearing of natural forests will release more carbon than will be sequestered by an equal area of afforestation in any given year.

emissions from ruminant livestock production systems. In 2000, for example, a Canadian power company and the Ugandan government initiated a 32-year project to improve cattle nutrition which would reduce methane emissions by an equivalent of 30 MtC while improving milk and meat productivity (Global Livestock Group 2000:1).

The realization that opportunities exist in agriculture to decrease carbon emissions has spurred new research on carbon emission reduction and sequestration potential related to land use and land use change. Preliminary findings indicate potentially large sequestration capacity. Houghton et al. (1999:576) estimate that farmer adoption of no tillage practices and government sponsored set-aside programs increased U.S. cropland carbon storage by some 138 MtC per year during the 1980s. Lal et al. (1998) estimate that the carbon storage of U.S. cropland could be further increased by some 120 to 270 MtC per year (7-19% of U.S. total emissions). And for the United Kingdom, Smith et al. (2000:8-9) calculated an increased carbon storage potential of about 10 MtC per year in the top 30 cm of cropland soils (about 7% of United Kingdom emissions). Globally, the IPCC (2000:184) estimates that feasible improvements in cropland management, grazing land management, agroforestry, and rice systems within existing land uses could increase carbon stocks by 125, 240, 26, and 7 MtC per year by 2010. Converting a greater share of land (around 126 Mha) into agroforestry could increase sequestration by some 390 MtC yr$^{-1}$.

There are real prospects for agriculture to play a greater role in reducing carbon emissions and increasing carbon sequestration. The primary cause for optimism is that the majority of ap-

proaches involved are also beneficial to agricultural productivity. However, evidence from low-income countries, in particular, is that farming communities face many obstacles in adopting improved practices, even when they know the potential benefits (Smith and Scherr forthcoming). There are also knowledge and data gaps associated with practically all the regional and global extrapolations underpinning the quantitative analysis, as well as problems in measuring and interpreting field data on carbon fluxes (see, for example, Duiker and Lal 2000). Criticisms have also been leveled at soil carbon sequestration analyses that overstate potential benefits by not fully accounting for the total flux of carbon associated with fertilizer production, irrigation, and the application of organic manure (Schlesinger 1999). Terrestrial biotic carbon sinks, some argue, could soon become saturated and would thus be vulnerable to future carbon release through land use changes brought about by new policies, land management practices, and climate change. This suggests that increasing sequestration in soil and vegetation sinks is only a short- to medium-term solution and that the only long-term solution is to reduce fossil fuel emissions (IPCC 1996b, *New Scientist* 1999).

## Summary of Indicators and Data

Some practical indicators of agroecosystems' contribution to global carbon services are the extents of crop- and pasture-land and the quantities of organic carbon stored in their soils (kilograms per square meter or tons per hectare), preferably to a depth of at least one meter. Not only is the soil carbon pool much greater than that of most agroecosystem vegetation, but it is also less vulnerable to loss, particularly from that proportion held below the plow layer. The soil carbon content indicator, which is relevant at multiple scales, is also straightforward to measure, although harmonization of data for different laboratory extraction and analysis methods is required. Although the strategic importance of routinely monitoring soil organic matter is being increasingly recognized as an indicator of soil productivity, soil degradation, and carbon services (ISRIC 1999; Smith et al. 1999), current data sets stem from an extremely sparse network of soil sampling points and, mostly, for a single point in time.

From a vegetation perspective, it is necessary to improve the description of land cover within the extent of agriculture and to improve the generic estimates of above- and below-ground carbon densities of field, pasture, and tree crops in different agroclimatic zones. Improving agricultural land cover estimates should be feasible through a judicious combination of remote sensing data and agricultural production statistics.

Given that agriculture-related burning contributes to both $CO_2$ and $CH_4$ emissions in some areas of the world, monitoring the incidence and extent of fires in such areas is both desirable and now feasible using satellite technology. There are significant practical difficulties, however, in reliably distinguishing between fires in agricultural and nonagricultural areas and between fires of natural and human origin.[25]

Monitoring trends in paddy rice areas, ruminant livestock, and total livestock would provide proxy indicators for different sources of methane emissions. Paddy rice areas are not explicitly compiled at an international level, but probably could readily be so through such institutions as FAO and IRRI and by improved discrimination in satellite-derived land cover data. To be most useful, such proxy measures would need to be augmented by hard-to-obtain technical information on nutrition quality, animal nutrition efficiency, and waste treatment management practices, among others. It is not obvious how this technical information could be compiled on a routine basis at affordable costs.

Although deforestation and reforestation rates are important in understanding the dynamics of land use, they are more specific indicators of forest carbon capacities. As long as the extent and location of agriculture is routinely monitored, applying the suggested indicators within that extent should provide adequate measures of the status of agroecosystem carbon storage services.

A potentially major source of information on the contribution of agroecosystems to global carbon services is related to the processes of the FCCC, as currently prescribed under the Kyoto Protocol. Because parties must document their carbon emission reduction activities, including land-use related offsets, and the quantities involved must be independently certified, there seems scope for gathering good quality, relevant information from this source at relatively low cost.

# ENDNOTES

1. For a list of countries included in each region see the report appendix at http://www.wri.org/wr2000 or http://www.ifpri.org.

2. For a similar comparison of FAO statistics and the IGBP classification of the satellite derived data, see Ramankutty and Foley 1998.

3. The land use statistics in the sections on Agricultural Land Use Balance and Trends, and Agricultural Land Use Dynamics came directly from FAOSTAT (1999) or were calculated by IFPRI based on FAOSTAT (1999) unless otherwise referenced.

4. For further maps and tables related to the input variables used to define agroclimatic zones, see the report appendix at http://www.wri.org/wr2000 or http://www.ifpri.org.

5. For characterizing the potential extent of irrigated agriculture, the moisture constraint is no longer valid and a separate, temperature-determined LGP has been calculated (IIASA 1999).

6. For the characterization of agroecosystems these areas were added to the satellite-defined extent of agriculture resulting in a percent of total land area of 28.6 and a percent of total population of 73.6 (see Table 5). These values are slightly higher than the percentages calculated within the agricultural extent based solely on satellite data (28.1 percent of total land area and 71.7 percent of total population).

7. An estimate of urban populations within agricultural areas based on the Stable Lights and Radiance Calibrated Lights of the World database (NOAA-NGDC 1998; Elvidge et al. 1997) drops the percentage of total population living within the extent of agriculture to about 50 percent of the world's total population (IFPRI calculation based on CIESIN 2000).

8. IFPRI calculated all statistics provided in the Food, Feed, and Fiber section based on 1961-97 data from FAOSTAT (1999) unless otherwise referenced.

9. Cropland was selected rather than agricultural land because pasture absorbs relatively little labor and pasture statistics are less reliable. Additionally, FAO has not reported pasture in its agricultural land use statistics, since 1994.

10. The 1995–97 average shares of cereals production for these regions (calculated in millions of 1989–91 dollars) are: East Asia, 25 percent; North America, 16 percent; Europe, 12 percent; Southeast Asia, 9 percent; Latin America, 6 percent; Former Soviet Union, 6 percent; and Sub-Saharan Africa, 4 percent.

11. The published agricultural GDP estimates used here include value added for fishery and forest products, in addition to crops and livestock.

12. These values are globally comparable because a single, international commodity price was used to weight local production of each commodity to obtain total value.

13. International food aid also represents a significant contribution to reducing food insecurity in Sub-Saharan Africa.

14. A soil mapping-unit can contain up to eight soil types, each occupying a known share of the unit. However, the *location* of each soil within the mapping unit is unknown. Where, as is normally the case, a single mapping unit spans multiple five-minute grids, it is assumed that each of the mapping-unit soils occurs within each grid cell according to the established area shares. If mapping-unit soils are not spatially heterogeneous, this assumption can be misleading. Across all grid cells in any given mapping unit, however, the interpretation will be correct.

15. For further tables related to the analysis of the overlay of the GLASOD map with the PAGE agricultural extent see the report appendix at www.wri.org or www.ifpri.org.

16. Soil organic matter is primarily carbon. The standard conversion factor is SOM is equal to 1.72 times soil organic carbon. But this ratio can vary according to the origin of the organic matter.

17. The data set description and the map are placed in the carbon section for consistency with the other PAGE reports.

18. Mesoamerica includes Costa Rica, El Salvador, Guatemala, Honduras, Mexico, Nicaragua, and Panama. The Caribbean includes Cuba, Dominican Republic, Haiti, and others. Andean countries include Bolivia, Colombia, Ecuador, Peru, and Venezuela. The Southern Cone includes Argentina, Brazil, Chile, Paraguay, and Uruguay.

19. It is important to distinguish the unsuitability of using these variables as short-term indicators of condition from their suitability for agroecosystem characterization, precisely because they are relatively stable.

20. For maps related to the overlay of irrigation data, see the report appendix at http://www.wri.org/wr2000 or http://www.ifpri.org.

21. IFPRI calculation based on IWMI (1999) and World Bank (2000).

22. The PAGE forest ecosystems report estimates carbon stocks based on latitude and the non-overlapping IGBP land cover categories (Matthews et al. 2000) hence reporting a lower estimate in sum for agricultural areas than is reported in this report for areas within the PAGE agricultural extent.

23. USGS EROS Data Center (EDC) provided a match for Olson's low and high estimates of carbon storage values for ecosystem complexes.

24. The PAGE analysis using the Soil Map of the World excludes Greenland and Antarctica

25. For further discussions of fires as an indicator of carbon sequestration capacity and biodiversity, see the PAGE forest and grassland ecosystems reports (Matthews et al. 2000; White et al. 2000).

# ABBREVIATIONS AND UNITS

| | |
|---|---|
| AEZ | Agroecological Zone |
| ASSOD | Assessment of Human-Induced Soil Degradation in South and Southeast Asia |
| CGIAR | Consultative Group on International Agricultural Research |
| $CH_4$ | Methane |
| CIAT | Centro Internacional de Agricultura Tropical |
| CIESIN | Consortium for International Earth Science Information Network |
| CIP | Centro Internacional de la Papa |
| $CO_2$ | Carbon Dioxide |
| DAD-IS | Domestic Animal Diversification Information System |
| DES | Dietary Energy Supply |
| DSE | Deutsche Stiftung für Internationale Entwicklung (German Foundation for International Development) |
| EDC | Earth Resources Observation Systems (EROS) Data Center |
| EPA | Environmental Protection Agency |
| EU | European Union |
| FAO | Food and Agriculture Organization of the United Nations |
| FCC | Fertility Capability Classification |
| FCCC | Framework Convention on Climate Change |
| FSU | Former Soviet Union |
| GATT | General Agreement on Tariffs and Trade |
| GCOS | Global Climate Observation System |
| GCTE | Global Change and Terrestrial Ecosystems |
| GDP | Gross Domestic Product |
| GEMS | Global Environment Monitoring System for Water |
| GHG | Greenhouse Gas |
| GLASOD | Global Assessment of Soil Degradation |
| GLCCD | Global Land Cover Characterization Database |
| GMO | Genetically Modified Organisms |
| GRDC | Global Runoff Data Centre |
| GtC | Gigaton ($1 \times 10^9$ tons) of Carbon |
| GTOS | Global Terrestrial Observing System |
| HYV | High Yielding Variety |
| IBSRAM | International Board for Soil, Research and Management |
| ICIS | International Crop Information System |
| ICSU | International Council of Scientific Unions |
| IFA | International Fertilizer Industry Association |
| IFDC | International Fertilizer Development Centre |
| IFPRI | International Food Policy Research Institute |
| IGBP | International Geosphere-Biosphere Programme |
| IIASA | International Institute for Applied Systems Analysis |
| ILEIA | Centre for Research and Information on Low External Input and Sustainable Agriculture |
| IPCC | Intergovernmental Panel on Climate Change |
| IPM | Integrated Pest Management |
| IPP | Intellectual Property Protection |
| IRRI | International Rice Research Institute |
| ISRIC | International Soil Reference Information Centre |
| ISSS | International Society of Soil Science |
| IUBS | International Union of Biological Sciences |
| IUCN | World Conservation Union |
| IWMI | International Water Management Institute |
| LAC | Latin America and the Caribbean |
| LGP | Length of Growing Period (in days) |
| LUCC | Land Use Cover and Change |
| LULUCF | Land Use, Land Use Change and Forestry |
| MASL | Meters Above Sea Level |
| MEA | Millennium Ecosystem Assessment |
| MFCAL | MultiFunctional Character of Agriculture and Land |
| Mha | Million Hectares |
| NGO | Nongovernmental Organization |
| NPK | Nitrogen (N), Phosphorus ($P_2O_5$), and Potassium ($K_2O$) |
| NSSL | National Seed Storage Library |
| NT | No Tillage Cultivation |
| OECD | Organisation for Economic Co-operation and Development |
| PAGE | Pilot Analysis of Global Ecosystems |
| ppb | Parts per billion (by volume) |
| pph | Persons per hectare |
| ppm | Parts per million (by volume) |
| SEI | Stockholm Environment Institute |
| SLCR | Seasonal Land Cover Region |
| SOM | Soil Organic Matter |
| SOTER | SOil and TERrain Database |
| SSA | Sub-Saharan Africa |
| SWNM | Soil, Water, and Nutrient Management Program |
| TAC | Technical Advisory Committee, CGIAR |
| TgC | Teragram ($1 \times 10^{12}$ grams) of Carbon |
| UEA | University of East Anglia |
| UNDP | United Nations Development Programme |
| UNEP | United Nations Environment Programme |
| UNESCO | United Nations Educational, Scientific and Cultural Organization |
| UNFCCC | United Nations Framework Convention on Climate Change |
| USDA | United States Department of Agriculture |
| USGS | United States Geological Survey |
| WANA | West Asia/North Africa |
| WCMC | World Conservation Monitoring Centre |
| WDI | World Development Indicators |
| WIEWS | World Information and Early Warning System on Plant Genetic Resources |
| WISE | World Inventory of Soil Emission Potentials |
| WMO | World Meteorological Organization |
| WOCAT | World Overview of Conservation Approaches and Technologies |
| WRI | World Resources Institute |
| WTO | World Trade Organization |
| WWF-US | World Wildlife Fund – United States |
| WWI | World Watch Institute |
| WYCOS | World Hydrological Cycle Observing System |

# REFERENCES

Alcamo, J., T. Henrichs and T. Rösch. 2000. *World Water in 2025-Global Modeling and Scenario Analysis for the World Commission on Water for the 21ˢᵗ Century*. Report A0002. Kassel, Germany: Center for Environmental Systems Research, University of Kassel.

Ali, M. and D. Byerlee. Forthcoming. "Productivity Growth and Resource Degradation in Pakistan's Punjab." In M. Bridges, F.P. de Vries, I. Hannam, R. Oldeman, S.J. Scherr and S. Sombatpanit, eds. *Responses to Land Degradation*. Enfield, NJ: Science Publishers.

Antle, J. and P. Pingali. 1994. "Pesticides, Productivity, and Farmer Health: A Philippine Case Study." *American Journal of Agricultural Economics* 76: 418-30.

Bathrick, D. 1998. "Fostering Global Well-Being: A New Paradigm to Revitalize Agricultural and Rural Development." 2020 Vision Food, Agriculture, and the Environment Discussion Paper No. 26. Washington, D.C.: IFPRI.

Batjes, N.H. 1996. "Total Carbon and Nitrogen in the Soils of the World." *European Journal of Soil Science* Vol. 47: 151-163.

Batjes, N.H. 1999. "Effects of Mapped Variation in Soil Conditions on Estimates of Soil Carbon and Nitrogen Stocks for South America." *Geoderma (submitted)*. Mimeo.

Batjes, N.H. 2000. Personal communication. September 14, 2000.

Batjes, N.H. and E.M. Bridges. 1994. "Potential Emissions of Radioactively Active Gases from Soil to Atmosphere with Special Reference to Methane: Development of a Global Database (WISE)." *Journal of Geophysical Research* 99 (D8): 16479-16489.

Bayer, C., J. Mielniczuk, T.J.C. Amado, L. Martin-Neto, S.V. Fernandes. 2000. "Organic Matter Storage in a Sandy Clay Loam Acrisol Affected by Tillage and Cropping Systems in Southern Brazil." *Soil and Tillage Research* 54: 101-109.

Bøjö, J. 1996. "The Costs of Land Degradation in Subsaharan Africa." *Ecological Economics* 16: 161-173.

Boserup, E. 1965. *The Conditions of Agricultural Growth: The Economics of Agrarian Change under Population Pressure*. Chicago: Aldine Publishing Co.

Bot, A.J. and F.O. Nachtergaele. 1999. "Physical Resource Potentials and Constraints at Regional and Country Level." AGLS Working Paper No. 7. Draft. Rome: FAO.

Brown, J. and T. Loveland. 1998. USGS/EROS Data Center (EDC). Personal communication.

Brown, L.R., M. Renner, B. Halweil. 1999. *Vital Signs 1999*. New York. W.W. Norton and Company.

Byerlee, D. 1996. "Modern Varieties, Productivity, and Sustainability: Recent Experience and Emerging Challenges." *World Development* Vol. 24, No. 4:697-718.

CIESIN. 2000. Center for International Earth Science Information Network (CIESIN) Columbia University, International Food Policy Research Institute (IFPRI), World Resources Institute (WRI), *Gridded Population of the World, Version 2 alpha*. Palisades, NY: CIESIN, Columbia University. Data available online at: http://sedac.ciesin.org/plue/gpw.

CGIAR. 2000. Consultative Group on International Agricultural Research. http://www.cgiar.org.

CIAT. 1996. Centro Internacional de Agricultura Tropical, Digital Map Dataset for Latin America and the Caribbean.

CIMMYT. 1998. Centro Internacional de Mejoramiento de Maiz y Trigo, *CIMMYT 1997/98 World Maize Facts and Trends Maize Production in Drought-Stressed Environments: Technical Options and Research Resource Allocation*. Heisey, P.W.; Edmeades, G.O. Mexico, DF (Mexico): CIMMYT.

CIP. 1999. Centro Internacional de la Papa, Potato Distribution dataset, personal communication with Robert Hijmans.

Conway, G. 1997. *The Doubly Green Revolution: Food for all in the 21ˢᵗ Century*. Ithaca: Cornell University Press.

Crissman, C., D. Cole, and F. Carpio. 1994. "Pesticide Use and Farm Worker Health in Ecuadorian Potato Production." *American Journal of Agricultural Economics* 76: 593-97.

DAD-IS. 2000. Domestic Animal Diversification Information System: http://dad.fao.org/.

Daily, G.C., P.R. Ehrlich, and G.R. Arturo Sánchez-Azofeifa. 2000. "Countryside Biogeography: Use of Human-Dominated Habitats by the Avifauna of Southern Costa Rica." *Ecological Applications* (in press).

DeFries, R.S., M.C. Hansen, J.R.G. Townshend, A.C. Janetos, and T.R. Loveland. 2000. "A New Global 1-Km Data Set of Percentage Tree Cover Derived from Remote Sensing." *Global Change Biology* 6: 247-254.

Delgado, C., M. Rosegrant, H. Steinfeld, S. Ehui, C. Courbois. 1999. "Livestock to 2020: The Next Food Revolution." Food, Agriculture, and the Environment Division Discussion Paper No. 28. Washington, D.C.: IFPRI.

DellaPenna, D. 1999. "Nutritional Genomics: Manipulating Plant Micronutrients to Improve Human Health." *Science* Vol. 285: 375-379.

Dewees, P. and S. J. Scherr. 1996. "Policies and Markets for Non-Timber Tree Products." Environment and Production Technology Division (EPTD) Discussion Paper No. 16. Washington, D.C.: International Food Policy Research Institute.

Dixon, R.K., J.K. Winjum and P.E. Schroeder. 1993. "Conservation and sequestration of carbon: The potential of forest and agroforest management practices." *Global Environmental Change* 159-173.

Döll, P. and S. Siebert. 1999. *A Digital Global Map of Irrigated Areas.* Report No. A9901. Centre for Environmental Systems Research. Germany: University of Kassel.

Dregne, H., M. Kassas, and B. Razanov. 1991. "A New Assessment of the World Status of Desertification." *Desertification Control Bulletin.* United Nations Environment Programme. 20: 6-18.

Duiker, W.W. and R. Lal. 2000. "Carbon Budget Study Using $CO^2$ Flux Measurements from a No-Till System in Central Ohio." *Soil and Tillage Research* Vol 54, 21-30.

Elvidge, C.D. et al. 1997. "Satellite Inventory of Human Settlements Using Nocturnal Radiation Emissions: a Contribution for the Global Toolchest." *Global Change Biology* 3 5: 387-396. View online at Nighttime Lights of the World Database: http://julius.ngdc.noaa.gov:8080/production/html/BIOMASS/night.html.

ESRI. 1996. Environmental Systems Research Institute, Inc., *World Countries 1995.* Included in ESRI Data and Maps. Volume 1. CD-ROM. Redlands, CA: ESRI. Country names and disputed territories updated at WRI and IFPRI as of 1999.

Eswaran, H., E. Van den Bergh, and P. Reich. 1993. "Organic Carbon in the Soils of the World." *Soil Science Society of America Journal* 57: 192-194.

Evans, L.T. 1998. *Feeding the Ten Billion: Plants and Population Growth.* Cambridge: Cambridge University Press.

Evenson, R.E., D. Gollin, and V. Santaniello, eds. 1998. *Agricultural Values of Plant Genetic Resources.* Wallingford: CABI Publishing.

FAO. 1978-81. Food and Agriculture Organization of the United Nations, *Report of the Agro-Ecological Zones Project.* World Soil Resources Report no. 48 vol. 1-4, Rome: FAO.

FAO. 1982. Food and Agriculture Organization of the United Nations, *Report of the Agro-Ecological Zones Project – Methodology and results for Africa.* Rome: FAO.

FAO. 1991. Food and Agriculture Organization of the United Nations, *The Digitized Soil Map of the World.* World Soil Resources Report 67/1 (Release 1).

FAO. 1993. Food and Agriculture Organization of the United Nations, *Food Production Yearbook.* Rome: FAO.

FAO. 1995. Food and Agriculture Organization of the United Nations, *Digital Soil Map of the World (DSMW) and Derived Soil Properties.* Version 3.5. CD-ROM.

FAO. 1996. Food and Agriculture Organization of the United Nations, *The Sixth World Food Survey.* Rome: FAO.

FAO. 1997a. Food and Agriculture Organization of the United Nation, Computer Printout of FAOSTAT's International Commodity Prices 1989-91. Personal Communication via Technical Advisory Committee, CGIAR. Rome: FAO.

FAO. 1997b. Food and Agriculture Organization of the United Nations, *State of the World's Forests 1997.* Rome: FAO, cited in WRI 1998: 185.

FAO. 1998. Food and Agriculture Organization of the United Nations, *The State of the World's Plant Genetic Resources for Food and Agriculture.* Rome: FAO.

FAO. 1999a. Food and Agriculture Organization of the United Nations, *The State of Food Insecurity in the World.* Rome: FAO. Available online at: http://www.fao.org/NEWS/1999/991004-e.htm.

FAO. 1999b. Food and Agriculture Organization of the United Nations, *Background Paper 6.* FAO/Netherlands Conference on the Multifunctional Character of Agriculture and Land. Rome: FAO. Online at: http://www.iisd.ca/linkages/sd/agr/index.html.

FAO. 1999c. Food and Agriculture Organization of the United Nations Land and Water Digital Media Series, *The Soil and Terrain Database for Latin America and the Caribbean.* CD-ROM. Rome: FAO.

FAO. 2000a. Food and Agriculture Organization of the United Nations, AQUASTAT program for Africa, Asia, Former Soviet Union & Near East. Online at: http://www.fao.org/ag/agl/aglw/aquastat/aquast1e.htm

FAO. 2000b. Food and Agriculture Organization of the United Nations, *International Undertaking on Plant Genetic Resources.* Online at: http://www.ext.grida.no/ggynet/agree/nat-con/fao-pgr.htm

FAO/IIASA. 1999. prerelease version of FAO/IIASA, Global Agro-ecological Zoning, 2000.

FAO/IIASA. 2000. Food and Agriculture Organization of the United Nations and International Institute for Applied Systems Analysis, Global Agro-ecological Zoning. FAO Land and Water Digital Media Series # 11. CD-ROM version forthcoming.

FAOSTAT. 1999. Food and Agriculture Organization of the United Nations, Statistical Databases. Online at: http://apps.fao.org.

FAOSTAT. 2000. Food and Agriculture Organization of the United Nations, Statistical Databases. Land Use Explanatory Notes. Online at: http://www.fao.org/waicent/faostat/agricult/landuse-e.htm. Agricultural Machinery Explanatory Note. Online at: http://www.fao.org/waicent/faostat/agricult/meansprod-e.htm.

Faurès, J.M. 2000. Food and Agriculture Organization of the United Nations, AGLW. Personal communication.

Fernandez-Cornejo, J. and S. Jans. 1999. *Pest Management in U.S. Agriculture.* Handbook No.717. Washington, D.C.: Resource Economics Division, ERS.

Fischer, G., K. Frohberg, M.L. Parry and C. Rosenzweig. 1996. "The Potential Effects of Climate Change on World Food Production and Security." In *Global Climate Change and Agricultural Production.* F. Bazzaz and W. Sombroek, eds. West Sussex: Wiley.

Fisher, M.J., I.M. Rao, M.A.Ayarza, C.E. Lascano, J.I. Sanz, R.J. Thomas, and R.R. Vera. 1994. "Carbon Storage by Introduced Deep-Rooted Grasses in the South American Savannas." *Nature* 371: 236-238.

Frye, W.W. 1984. "Energy Requirement in No-tillage." p.127-151 In. R.E. Phillips and S.H Phillips, eds. *No Tillage Agricultural Principles and Practices*. New York: Van Nostrand Rheinhold. Cited in Kern and Johnson 1993:204.

Gallopin, G. and F. Rijsberman. 1999. "Three Global Water Scenarios." *Report to the World Commission on Water for the 21st Century*. Stockholm Environment Institute.

GCOS. 2000. Global Climate Observing System. Online at: http://www.wmo.ch/web/gcos/.

GEMS/Water. 2000. United Nations Environment Programme Global Environment Monitoring System Freshwater Quality Programme. Online at: http://www.cciw.ca/gems/.

Ghassemi, F., A.J. Jakeman, H.A. Nix. 1995. *Salinisation of Land and Water Resources: Human Causes, Extent, Management, and Case Studies*. Wallingford: CAB International.

Gleick, P.H. (ed.). 1993. *Water in Crisis: A Guide to the World's Fresh Water Resources*. Oxford: Oxford University Press.

Global Land Cover Characteristics Database (GLCCD), Version 1.2. 1998. Loveland, T.R., B.C. Reed, J.F. Brown, D.O. Ohlen, Z. Zhu, L. Yang, and J.W. Merchant. 2000. "Development of a Global Land Cover Characteristics Database and IGBP DISCover from 1 km AVHRR Data." *International Journal of Remote Sensing* 21 (6/7): 1303-1330). Data available online at: http://edcdaac.usgs.gov/glcc/glcc.html.

Global Livestock Group. 2000. "The GLG and a Canadian Electric Utility Company Sign 30 Million Ton CO$^2$ Offset Project Agreement." Online at: http://www.theglg.com/newsletter1662/newsletter_show.htm?doc_id=26778

Goklany, I.M. 2000. "Richer is More Resilient: Dealing with Climate Change and More Urgent Environmental Problems." *Earth Report 2000*. R. Bailey (ed.)

Gould, F. and M.B. Cohen. 2000. "Sustainable Use of Genetically Modified Crops in Developing Countries." In G.J. Persley and M.M. Lantin, (eds.) *Agricultural Biotechnology and the Poor*. 2000: 139-148.

GRDC. 2000. Global Runoff Data Center. Online at: http://www.bafg.de/grdc.htm.

Gregory, P.J., J.S.I. Ingram, B. Campbell, J. Goudriaan, L.A. Hunt, J.J. Landsberg, S.Linder, M. Stafford Smith, R.W. Sutherst and C. Valentin. 1999. *Managed Production Systems*. In B. Walker, W. Steffen, J. Canadell and J. Ingram (eds). The Terrestrial Biosphere and Global Change: Implications for Natural and Managed Ecosystems. Cambridge: Cambridge University Press: 229-270.

Harris, D.R. and G.C. Hillman (eds). 1989. *Foraging and Farming: The Evolution of Plant Exploitation*. London: Unwin Hyman.

Harris, G. 1998. "An Analysis of Global Fertilizer Application Rates for Major Crops." Paper presented at the Agro-Economics Committee Fertilizer Demand Meeting at the IFA Annual Conference, Toronto. May 1998.

The Heinz Center. 1999. Designing a Report on the State of the Nations Ecosystems: Selected measurements for Croplands, Forests and Coasts & Oceans. The H. John Heinz III Center, Washington, D.C. Online at: www.us-ecosystems.org.

Heisey, P.W., and G.O. Edmeades. 1999. *World Maize Facts and Trends 1997-98; Maize Production in Drought-Stressed Environments: Technical Options and Research Resource Allocation."* Mexico D.F.: International Maize and Wheat Improvement Center (CIMMYT).

Heisey, P.W., M. Lantican and J. Dubin. 1999. Assessing the Benefits of Internatioal Wheat Breeding Research: An Overview of the Global Wheat Impacts Study. Part 2, *CIMMYT 1998-99 World Wheat Facts and Trends*. Mexico, D.F.: (CIMMYT).

Henao, J. 1999. "Assessment of Plant Nutrient Fluxes and Gross Balances in Soils of Agricultural Lands in Latin America." Report and data prepared as part of the Pilot Assessment of Global Ecosystems (PAGE). Alabama: IFDC (International Fertilizer Development Center).

Henao, J. and C. Baanante. 1999. *Estimating Rates of Nutrient Depletion in Soils of Agricultural Lands of Africa*. Technical Bulletin International Fertilizer Development Center (IFDC)-T-48. Muscle Shoals, Alabama: IFDC.

Houghton, J. 1997. *Global Warming: The Complete Briefing*. Cambridge: Cambridge University Press.

Houghton, R.A., and J.L. Hackler. 1995. "Continental Scale Estimates of the Biotic Carbon Flux from Land Cover Change: 1850 to 1980." Report NDP-050. Online at: http://cdiac.esd.ornl.gov/ndps/ndp050.html

Houghton, R.A., J.L. Hackler, and K.T. Lawrence. 1999. "The U.S. Carbon Budget: Contributions from Land Use Change." *Science* 285: 574-578.

Huang, J. 1996. "Technological Change: Rediscovering the Engine of Productivity in China's Rural Economy." *Journal of Development Economics*. 49: 337-369.

ICIS. 2000. International Crop Information System http://www.cgiar.org/icis/homepage.htm

ICLARM. 1999. *The Legal and Institutional Framework, and Economic Valuation of Resources and Environment in the Mekong River Region: A Wetlands Approach*. The Philippines: International Center for Living Aquatic Resources Management.

IFDC/IFA/FAO. 1997. "Survey of Fertilizer Rates of Use on the Individual Major World Crops." Cited in Harris, "An Analysis of Global Fertilizer Application Rates for Major Crops." 1998.

IGBP. 1998. International Geosphere Biosphere Programme (IGBP) Data and Information System, IGBP-DIS Global 1-km Land Cover Set DISCover.

IGBP. 2000. International Geosphere Biosphere Programme. Land Use and Land Cover Change Program. Online at: http://www.igbp.kva.se/lucc.html

IIASA. 1999. International Institute for Applied Systems Analysis. Agroecosystem Component of the Pilot Analysis of Global Ecosystems (PAGE) Digital Global Datasets and Publication Contribution: A Summary of the Databases Provided to WRI from the FAO/IIASA Global Agro-Ecological Zones Assessment.

IMF. 1987. International Monetary Fund, International Financial Statistics Yearbook 1987.

IMF. 1998. International Monetary Fund, International Financial Statistics Yearbook 1998.

IPCC. 1996a. Intergovernmental Panel on Climate Change, *Climate Change 1995: The Science of Climate Change*. Cambridge: Cambridge University Press.

IPCC. 1996b. Intergovernmental Panel on Climate Change, *Climate Change 1995: Impacts, Adaptations and Mitigation of Climate Change: Scientific-Technical Analyses*. Cambridge: Cambridge University Press.

IPCC. 2000. Intergovernmental Panel on Climate Change, *Land Use, Land-Use Change, and Forestry*. R.T. Watson, I.R. Noble, B. Bolin, N.H. Ravindranath, D.J. Verardo, D.J. Dokken (eds.). Cambridge: Cambridge University Press. Summary available online at: http://www.ipcc.ch/pub/SPM_SRLULUCF.pdf

IRRI. 2000. International Rice Research Institute, *Something to Laugh About*. Los Banos: IRRI.

Ismail, I., R.L Blevins, and W.W. Frye. 1994. *Soil Science Society of America Journal* Vol 58. 193. Cited in Schlesinger 1999:2095.

ISRIC. 1999. International Soil Reference Information Centre Documentation on LAC Soil Carbon map prepared for IFPRI/WRI as part of the PAGE project.

IUCN. 1996. World Conservation Union IUCN Red List of Threatened Animals. IUCN, Gland, Switzerland.

IWMI. 1999. International Water Management Institute, *World Water Vision Scenarios*.

James, C. 1999. "Global Review of Commercialized Transgenic Crops." ISAAA Brief. Ithaca, New York: ISAAA.

Johnson, B. 2000. "Genetically Modified Crops and Other Organisms: Implications for Agricultural Sustainability and Biodiversity." In G.J. Persely and M.M. Lantin, Eds. *Agricultural Biotechnology and the Poor: Proceedings of an International Conference*, Washington, D.C. 21-22 October 1999. Consultative Group on International Agricultural Research, Washington, D.C., 131-138.

Kern, J.S. and M.G. Johnson. 1993. "Conservation Tillage Impacts on National Soil and Atmospheric Carbon Levels." *Soil Science Society of America Journal* 57:200-210.

Kindt, R., S. Muasya, J. Kimotho, A. Waruhiu. 1997. "Tree Seed Suppliers Director—Sources of Seeds and Microsymbionts." Nairobi: ICRAF.

Klohn, W.E. and B.G. Appelgren. 1998. "Challenges in the Field of Water Resource Management in Agriculture." In *Sustainable Management of Water in Agriculture: Issues and Policies*. OECD Proceedings. Paris: OECD, 33.

Lal, R. 1995. "Erosion-Crop Productivity Relationships for Soil of Africa." *Soil Science Society of America Journal* 59 (3): 661-667.

Lal, R. 2000. "Soil Management in the Developing Countries". *Soil Science* 165. No 1: 57-72.

Lal, R., J.M Kimble, R.F. Follett, and C.V. Cole. 1998. *The Potential of U.S. Cropland to Sequester Carbon and Mitigate the Greenhouse Effect*. Chelsea: Ann Arbor Press. Cited in Rosenweig and Hillel. "Soils and Global Climate Change: Challenges and Opportunities." 2000:54.

Leakey, R.R.B. 1999. *Win:Win Landuse Strategies for Africa: Matching Economic Development with Environmental Benefits through Tree Crops*. Paper presented to the USAID Sustainable Tree Crop Development Workshop "Strengthening Africa's Competitive Position in Global Markets", Washington 19-21. October.

Lefroy, E.C., R.J. Hobbs, M.H. O'Connor and J.S. Pate. 1999. "Special Issue: Agriculture as a Mimic of Natural Ecosystems." *Agroforestry Systems* 45 (1-3).

Lindert, P. forthcoming 2001. "Soil Change and Agriculture in Two Developing Countries." Chapter 8 in *Agricultural Science Policy: Changing Global Agendas*. eds. J.M. Alston, P.G. Pardey, and M.J. Taylor. Baltimore: Johns Hopkins University Press.

Loveland, T.R., B.C. Reed, J.F. Brown, D.O. Ohlen, Z. Zhu, L. Yang, and J.W. Merchant. 2000. "Development of a Global Land Cover Characteristics Database and IGBP DISCover from 1 km AVHRR Data." *International Journal of Remote Sensing* 21 (6/7): 1303-1330.

MAFF. 2000. Ministry of Agriculture, Fisheries and Food, *Agri-environment Schemes*. Online at: http://www.maff.gov.uk/environ/envindx.htm

Mann. 1999. "Crop Scientists Seek a New Revolution." *Science* Vol. 283: 310-314.

Matthews, E., S. Murray, R. Payne and M. Rohweder. 2000. *Pilot Analysis of Global Ecosystems: Forest Ecosystems*. Washington, D.C.: World Resources Institute.

McCalla, A. 1999. "Rural Development: Celebrating Achievements and Looking at Future Opportunities." Presented to the Environmentally and Socially Sustainable Development Group, The World Bank, Washington, D.C. December.

McNeely, J. and S.J. Scherr. Forthcoming. *Saving our Species: How the Earth's Biodiversity Depends on Progress in Agriculture*. Washington, D.C.: Future Harvest.

Mew, T. 2000. *Research Initiatives in Cross Ecosystem: Exploiting Biodiversity for Pest Management*. Los Banos: International Rice Research Institute.

Molden, D.J., R. Sakthivadivel, C.J. Perry, and C. de Fraiture. 1998. "Indicators for Comparing Performance of Irrigated Agricultural Systems." Research Report No 20. Colombo, Sri Lanka: International Water Management Institute.

Morris, M.L. and P.W. Heisey. 1998. "Achieving Desirable Levels of Crop Diversity in Farmers' Fields: Factors Affecting the Production and Use of Commercial Seed." In *Farmers, Gene Banks and Crop Breeding: Economic Analyses of Diversity in Wheat, Maize, and Rice.* M. Smale (ed.): 217-237.

NOAA-NGDC. 1998. National Oceanic and Atmospheric Administration-National Geophysical Data Center, *Stable Lights and Radiance Calibrated Lights of the World CD-ROM.* Boulder, CO: NOAA-NGDC. View Nighttime Lights of the World database online at: http://julius.ngdc.noaa.gov8080/production/html/BIOMASS/night.html.

Nelson, M. and M. Maredia. 1999. "Environmental Impacts of the CGIAR: An Initial Assessment." Rome: Impact Assessment and Evaluation Group Secretariat, Consultative Group for International Agricultural Research. Online at: http://www.inrm.cgiar.org/documents/IAEG/iaegdocument.htm.

*New Scientist.* 1999. "This Week." 23 October: 20.

Oerke, E.C., H.W. Dehne, F. Schonbeck and A. Weber. 1994. *Crop Production and Crop Protection.* Amsterdam: Elsevier.

Oldeman, L.R. 1994. "The Global Extent of Soil Degradation." In Greenland, D.J. and I. Szabolcs (eds). *Soil Resilience and Sustainable Land Use.* Wallingford: CAB International, 99-118.

Oldeman L.R. 1998. "Soil Degradation: A Threat to Food Security?" Report 98/01. Wageningen: International Soil Reference and Information Centre.

Oldeman. L.R., R.T.A. Hakkeling, and W.G. Sombroek. 1991a. *World Map of the Status of Human Induced Soil Degradation: an Explanatory Note. Second revised edition.* Wageningen and Nairobi: International Soil Reference and Information Centre (ISRIC) and United Nations Environmental Programme (UNEP).

Oldeman, L.R., van Engelen, V.W.P. and Pulles, J.H.M. 1991b. "The Extent of Human-Induced Soil Degradation." Appendix to Oldeman, L.R., Hakkeling, R.T.A. and Sombroek, W.G. *World Map of the Status of Human-Induced Soil Degradation: An Explanatory Note.* Wageningen: International Soil Reference and Information Centre (ISRIC): 27-33.

Olson, D.M., et al. 1999. Terrestrial Ecoregions of the World. Unpublished data. Washington, D.C: World Wildlife Fund – United States (WWF-U.S.).

Olson, J.S., J.A. Watts, and L.J. Allison. 1983. "Carbon on Live Vegetation of Major World Ecosystems." Report ORNL-5862. Oak Ridge, Tennessee: Oak Ridge National Laboratory.

Persley, G.J. 2000. "Agricultural Biotechnology and the Poor: Promethean Science." In G.J. Persley and M.M. Lantin (eds) *Agricultural Biotechnology and the Poor.* Proceedings of an International Conference, Washington, D.C., 21-22 October 1999. Consultative Group on International Agricultural Research, Washington, D.C., 3-24.

Persley, G.J. and M.M. Lantin, (eds.) 2000. *Agricultural Biotechnology and the Poor.* Proceedings of an International Conference, Washington, D.C., 21-22 October 1999. Consultative Group on International Agricultural Research, Washington, D.C., 139-148.

Pingali, P.L., and P.W. Heisey. 1999. "Cereal Crop Productivity in Developing Countries: Past Trends and Future Prospects." CIMMYT Economics Program Working Paper No. 99-03.

Pinstrup-Andersen, P., R. Pandya-Lorch and M. Rosegrant. 1999. "World Food Prospects: Critical Issues for the Early Twenty-First Century." 2020 Food Policy Report. Washington, D.C.: IFPRI.

Postel, S. 1993. "Water and Agriculture." Chapter in Gleick, P.H. (ed.), *Water in Crisis: A Guide to the World's Fresh Water Resources.* Oxford: Oxford University Press, 56-66.

Postel, S. 1997. *The Last Oasis: Facing Water Scarcity.* Washington, D.C.: Worldwatch Institute.

Postel, S. 1999. *Pillar of Sand.* New York: W. W. Norton & Co.

Postel, S., G.C. Daily, and P.R. Ehrlich. 1996. "Human Appropriation of Renewable Fresh Water." *Science* 271: 785-788. (February 9)

Prince, S.D., E. Brown de Colstoun and L.L. Kravitz. 1998. "Evidence from Rain-Use Efficiencies Does Not Indicate Extensive Sahelian Desertification." *Global Change Biology* 4:101-116.

Ramankutty, N. and J.A. Foley. 1998. "Characterizing patterns of global land use: an analysis of global croplands data." *Global Biogeochemical Cycles* 12(4):667–685.

Reijntjes, C., M. Minderhoud-Jones, and P. Laban. 1999. *Leisa in Perspective: 15 Years ILEIA.* Leusden: ILEIA (Institute for Low External Input and Sustainable Agriculture).

Revenga, C., J. Brunner, N. Henninger, K. Kassem, R. Payne. 2000. *Pilot Analysis of Global Ecosystems: Freshwater Systems.* Washington, D.C.: World Resources Institute.

Ricketts, T.R., G.C. Daily, P.R. Ehrlich, and J.P. Fay. 2000. "Countryside Biogeography of Moths in a Fragmented Landscape: Biodiversity in Native and Agricultural Habitats." *Conservation Biology* (in press).

Rosegrant, M. 1997. "Water Resources in the Twenty-First Century: Challenges and Implications for Action." Environment and Production Technology Division Discussion Paper No. 20. Washington, D.C.: IFPRI.

Rosegrant, M. and C. Ringler. 1997. "World Food Markets into the 21st Century: Environmental and Resource Constraints and Policies." *The Australian Journal of Agricultural and Resource Economics* Vol. 41, No. 3.

Rosegrant, M. and C. Ringler. 2000. "Impact on Food Security and Rural Development of Reallocating Water from Agriculture." *Water Policy*: 567-586.

Rosenzweig, C. and D. Hillel. 2000. "Soils and Global Climate Change: Challenges and Opportunities." *Soil Science* 165. No 1: 57-72.

Sanchez, P.A., W. Couto, and S.W. Buol. 1982. "The Fertility Capability Soil Classification System: Interpretation, Application and Modification." *Geoderma* Vol 27(4): 283-309.

Scherr, S.J. 1995. "Economic Factors in Farmer Adoption of Agroforestry: Patterns Observed in Western Kenya." *World Development* 23 (5): 787-804.

Scherr, S.J. 1999a. "Poverty-Environment Interactions in Agriculture: Key Factors and Policy Implications." Poverty and Environment Initiative Paper No.3. United Nations Development Programme and the European Commission: New York.

Scherr, S.J. 1999b. "Soil Degradation: A Threat to Developing-Country Food Security?" 2020 Vision Food, Agriculture, and the Environment Discussion Paper No. 27. Washington, D.C.: International Food Policy Research Institute.

Scherr, S.J. 2000. "A Downward Spiral? Research Evidence on the Relationship Between Poverty and Environment." *Food Policy.* Vol. 25 (4): 479-498.

Scherr, S.J. and S. Yadav. 1995. "Land Degradation in Developing World: Implications for Food, Agriculture, and the Environment." 2020 Vision, Food, Agriculture, and the Environment Discussion Paper No. 14. Washington, D.C.: IFPRI.

Schlesinger, W.H. 1990. "Evidence From Chronosequence Studies for a Low Carbon-Storage Potential of Soils." *Nature* Vol. 348:232-234. Cited in Kern and Johnson "Conservation Tillage Impacts on National Soil and Atmospheric Carbon Levels." 1993:209.

Schlesinger, W.H. 1999. "Carbon Sequestration in Soils." *Science* 284: 2095.

Sebastian, K. and S. Wood. 2000. "Spatial Aspects of Evaluating Technical Change in Agriculture in Latin America and the Caribbean (LAC)." Paper prepared as part of the BID-sponsored project "Policies on Food, Agriculture and the Environment, Indicators and Priorities for Agricultural Research."

Seckler, D., U. Amarasinghe, D. Molden, R. de Silva, and Barker, R. 1998. "World Water Demand and Supply, 1990 to 2025: Scenarios and Issues." Research Report No 19. International Water Management Institute. Colombo, Sri Lanka.

Seré, C. and H. Steinfeld. 1996. World Livestock Production Systems: Current Status, Issues and Trends. FAO Animal Production and Health Paper 127. Rome: FAO.

SFI. 2000. Soil Fertility Initiative. Online at: http://www.fertilizer.org/PUBLISH/sfi.htm

Shand, H. 1997. *Human Nature: Agricultural Biodiversity and Farm-based Food Security.* Ottawa, Ontario: Rural Advancement Foundation International.

Shiklomanov, I.A. 1996. "Assessment of Water Resources and Water Availability in the World." Report for the Comprehensive Global Freshwater Assessment of the United Nations. State Hydrological Institute, St. Petersburg, Russia (February draft).

Smale, M. 2000."Economic Incentives for Conserving Crop Genetic Diversity on Farms: Issues and Evidence." Plenary Session paper presented at the XXIV International Conference of Agricultural Economists, August 13-19, Berlin, Germany.

Smaling, E.M.A., S.M. Nandwa, and B.H. Janssen. 1997. "Soil Fertility in Africa is at Stake." In *Replenishing Soil Fertility in Africa.* Soil Science Society of America Special publication Number 41, edited by R. Buresh, P.A. Sanchez, and F. Calhoun, p.47-62. Madison: American Society of Agronomy.

Smith, B.D. 1998. *The Emergence of Agriculture.* New York: Scientific American Library.

Smith, C.W. 1989. "The Fertility Capability Classification System (FCC) – 3rd Approximation: A Technical Soil Classification System Relating Pedon Characterization Data to Inherent Fertility Characteristics." Doctoral dissertation. North Carolina State University.

Smith, C.W., S.W. Buol, P.A. Sanchez and R.S. Yost. 1997. *Soil Fertility Capability Classification System (FCC), 3rd Approximation: The Link Between Pedologist and Soil Fertility Manager.*

Smith, J. and S.J. Scherr. Forthcoming. "Capturing the Value of Forest Carbon for Local Livelihoods." Policy Report. Centre for International Forestry Research and Forest Trends, Bogor, Indonesia.

Smith, P., D.S. Powlson, P. Falloon and J.U. Smith. 1999. "SOMNET. A Global Network and Database of Soil Organic Matter Models and Long-Term Experimental Datasets." from the Global Change and Terrestrial Ecosystems Soil Organic Matter Network (GCTE-SOMNET) webpage. Available at: http://www.iacr.bbsrc.ac.uk/res/depts./soils/somnet/tintro.html

Smith, P., D.S. Powlson, J.U. Smith, P. Falloon and K. Coleman. 2000. "Meeting the UK's Climate Change Commitments: Options for Carbon Mitigation on Agricultural Land." *Soil Use and Management* Vol. 16: 1-11.

Smith, W. et al. 2000. *Canada's Forests at a Crossroads: An Assessment in the Year 2000.* A Global Forest Watch Canada Report. Washington, D.C.: World Resources Institute.

Sombroek, W. G. and R. Gommes. 1996. "The Climate Change-Agriculture Conundrum." In *Global Climate Change and Agricultural Production.* F. Bazzaz and W. Sombroek, eds. West Sussex: Wiley.

Stern, D.I., and R.K. Kaufmann. 1998. "Annual Estimates of Global Anthropogenic Methane Emissions: 1860-1994." *Trends Online: A Compendium of Data on Global Change.* Carbon Dioxide Information Analysis Center, Oak Ridge National Laboratory, U.S. Department of Energy, Oak Ridge, Tenn., U.S.A. http://cdiac.esd.ornl.gov/trends/meth/ch4.htm - livestock

Stewart, B.A., R. Lal, and S.A. El-Swaify. 1991. "Sustaining the Resource Base of an Expanding World Agriculture." In *Soil Management for Sustainability,* edited by R. Lal and F.J. Pierce, pp. 125-44.

Stoorvogel, J.J. and E.M.A. Smaling. 1990. *Assessment of Soil Nutrient Depletion in Sub-Saharan Africa: 1983-2000* (Three Volumes). Wageningen: ISRIC.

Subramanian, A., N.V. Jagannathan, and R. Meinzen-Dick. 1997. "User Organizations for Sustainable Water Services." Technical Paper 354. Washington, D.C.: World Bank.

Swift, M.J. and J.M. Anderson. 1999. "Biodiversity and Ecosystem Function in Agricultural Systems." In *Biodiversity and Ecosystem Function*. Schulze, E-D. and H.A. Mooney (eds). Berlin: Springer-Verlag.

SWNM. 2000. Soil Water and Nutrient Management Program, Online at: http://198.93.225.137/projects/soils/swnm.htm

Thrupp, L. A. 1998. *Cultivating Diversity: Agrobiodiversity and Food Security*. Washington D.C.: World Resources Institute.

Tiessen, H., E. Cuevas, and P. Chacon. 1994. "The Role of Soil Organic Matter in Sustaining Soil Fertility." *Nature* 371: 783-785.

Tucker, C. J. and John R. G. Townshend. 2000. "Strategies for Monitoring Tropical Deforestation Using Satellite Data" *International Journal of Remote Sensing* Vol. 21, No. 6:1461-1472.

Turner II, B.L., W.C. Clark, R.W. Kates, J.F. Richards, J.T Mathews and W.B. Meyer. (eds). 1990. *The Earth as Transformed by Human Action: Global and Regional Change in the Biosphere over the Past 300 Years*. Cambridge: Cambridge University Press.

UN. 1997. United Nations, *Comprehensive Assessment of the Freshwater Resources of the World*. Stockholm: Commission for Sustainable Development, Stockholm Environment Institute: section 37.

UN. 2000. Official web site of the United Nations Framework Convention on Climate Change (FCCC). Online at: http://www.unfccc.de/

UNCCD. 2000. United Nations Convention to Combat Desertification. Online at: http://www.unccd.int/main.php

UNEP. 1995. United Nations Environment Programme, *Global Biodiversity Assessment*. V.H. Heywood (ed.). Cambridge: Cambridge University Press.

UNFCCC. 2000. United Nations Framework Convention on Climate Change, Land Use, Land Use Change and Forestry (LULUCF). Special Report to the Inter-Governmental Panel on Climate Change.

UNH-GRDC. 1999. University of New Hampshire – Global Runoff Data Center, World Basin Database. Available online at: http://www.grdc.sr.unh.edu/html/Data/index.html.

UEA. 1998. University of East Anglia Climatic Research Unit. http://www.cru.uea.ac.uk/cru/cru.htm

University of Kassel. 1999. A Digital Global Map of Irrigated Areas.

University of Maryland. 1999. UMD Department of Geography, Rain Use Efficiency and Net Primary Production Database prepared for WRI as part of the PAGE project.

USDA. 1994. United Stated Department of Agriculture, *Major World Crop Areas and Climatic Profiles. World Agricultural Outlook Board, U.S. Department of Agriculture*. Agricultural Handbook No. 664. Online version available at http://www.usda.gov/oce/waob/jawf/profiles/mwcacp2.htm.

USDA. 2000. United Stated Department of Agriculture, *Farm Income and Financial Conditions*. Available online at: http://www.ers.usda.gov/Briefing/baseline/1999/income.htm#income.

USDA-NASS. 1999. Historical Track Records for Commodities. Available online at: http://www.usda.gov/nass/pubs/histdata.htm. Wheat data downloaded 13-Sept, 1999.

USGS. 1998. United States Geological Surveys Earth Resources Observation Systems (EROS) Data Center, *GTOPO30: Global 30 Arc Second Elevation Data Set*. Sioux Falls, SD: USGS EDC.

USGS. 1999. United States Geological Surveys, *The Quality of Our Nation's Waters-Nutrients and Pesticides*. U.S. Geological Survey Circular 1225. 82 p.

USGS EDC. 1999a. United States Geological Surveys Earth Resources Observation Systems (EROS) Data Center, *1km Land Cover Characterization Database revisions for Latin America*. Sioux Falls, SD: USGS EDC. Unpublished data.

USGS EDC. 1999b. United States Geological Surveys Earth Resources Observation Systems (EROS) Data Center, *Carbon in Live Vegetation Map*. Sioux Falls, SD: USGS EDC. Unpublished data. For this digital map, USGS EDC applied carbon density numbers from a previous study (Olson et al. 1983) to a more recent global vegetation map (GLCCD 1998).

van de Fliert, E. 1993. "Integrated Pest Management: Farmer Field Schools Generate Sustainable Practices. A Case Study in Central Java Evaluating IPM Training." Doctoral Dissertation. Wageningen Agricultural University Papers 93-3.

van Lynden, G.W.J. and L.R. Oldeman. 1997. "The Assessment of the Status of Human-Induced Soil Degradation in South and South-East Asia." United Nations Environment Programme (UNEP), Food and Agriculture Organization of the United Nations (FAO), International Soil Reference and Information Centre (ISRIC), Wageningen, The Netherlands.

WCMC. 1999. World Conservation Monitoring Centre, *Protected Areas Database*. Cambridge: United Kingdom: WCMC. Unpublished data.

White, R., S. Murray, M. Rohweder. 2000. *Pilot Analysis of Global Ecosystems: Grassland Ecosystems*. Washington, D.C.: World Resources Institute.

WIEWS. 2000. World Information and Early Warning System on Plant Genetic Resources. Online at: http://apps2.fao.org/wiews/index.html.

Wilson, E.O. 1992. *The Diversity of Life*. Cambridge: Harvard University Press.

Winograd, M. and Farrow, A. 1999. Agroecosystem Assessment for Latin America. Prepared for WRI as part of the PAGE project.

World Bank. 2000. *World Development Indicators*. Washington, D.C.: World Bank.

WMO. 1997. World Meteorological Association, *Comprehensive Assessment of the Freshwater Resources of the World*. Geneva: World Meteorological Association.

WRI. 1998. World Resources Institute, *World Resources 1998-99: A Guide to the Global Environment.* Washington, D.C.: World Resources Institute.

WRI. 2000. World Resources Institute, *World Resources 2000-2001: People and Ecosystems.* Washington, D.C.: World Resources Institute.

WWF. 2000. World Water Forum. Online at: http://www.worldwaterforum.org/.

WYCOS. 2000. Official webpage for the World Hydrological Cycle Observing System. Online at: http://www.wmo.ch/web/homs/whycos.html.

Young, A. 1998. *Land Resources: Now and for the Future.* Cambridge: Cambridge University Press, 57-62.

Yudelman, M., A. Ratta, and D. Nygaard. 1998. "Pest Management and Food Production: Looking to the Future." 2020 Vision Food, Agriculture, and the Environment Discussion Paper 25. Washington, D.C.: IFPRI.

## Map 1
# PAGE Agricultural Extent

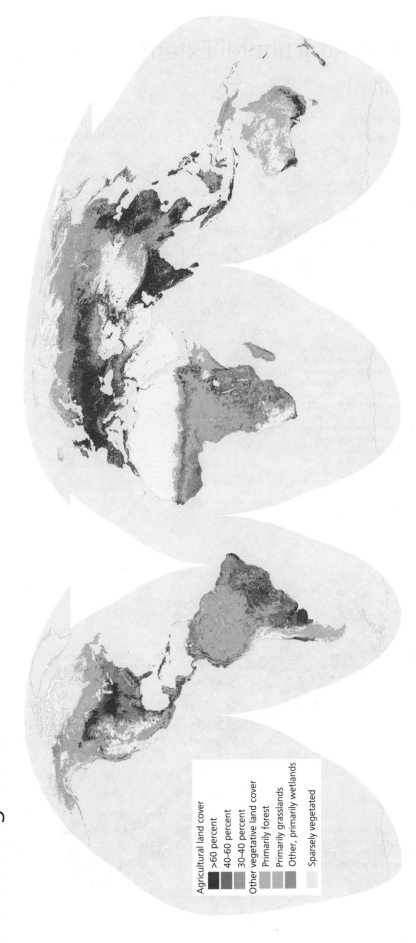

**Agricultural land cover**
- >60 percent
- 40-60 percent
- 30-40 percent

**Other vegetative land cover**
- Primarily forest
- Primarily grasslands
- Other, primarily wetlands

- Sparsely vegetated

**Source:** IFPRI reinterpretation of GLCCD 1998; USGS EDC 1999a.

**Projection:** Interrupted Goode's Homolosine

**Note:** Other vegetative land cover might contain as much as 30 percent agricultural land, but the actual amount cannot be determined using the GLCCD dataset. The satellite-derived estimate of agricultural extent is likely to under-represent some types of agricultural land cover including: extensive dryland arable farming, pastures, irrigated areas, and permanent crops - particularly in forest margins. Since the satellite interpretation was perfomed on a regional basis, the nature and extent of this under-reporting varies among regions.

Agroecosystems

# Map 2

# Comparison of Agricultural Extent for Central America

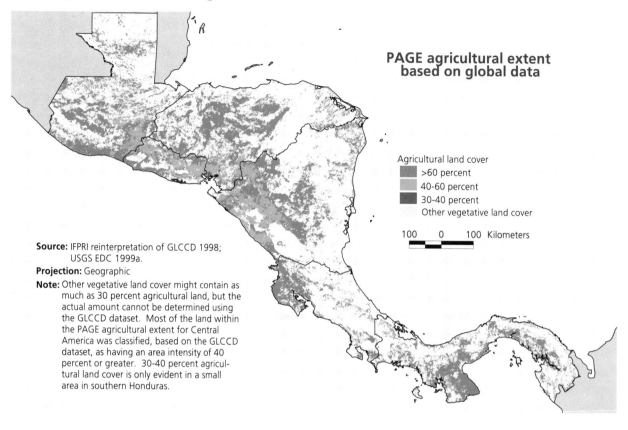

**PAGE agricultural extent based on global data**

Agricultural land cover
- >60 percent
- 40-60 percent
- 30-40 percent
- Other vegetative land cover

100   0   100   Kilometers

**Source:** IFPRI reinterpretation of GLCCD 1998; USGS EDC 1999a.

**Projection:** Geographic

**Note:** Other vegetative land cover might contain as much as 30 percent agricultural land, but the actual amount cannot be determined using the GLCCD dataset. Most of the land within the PAGE agricultural extent for Central America was classified, based on the GLCCD dataset, as having an area intensity of 40 percent or greater. 30-40 percent agricultural land cover is only evident in a small area in southern Honduras.

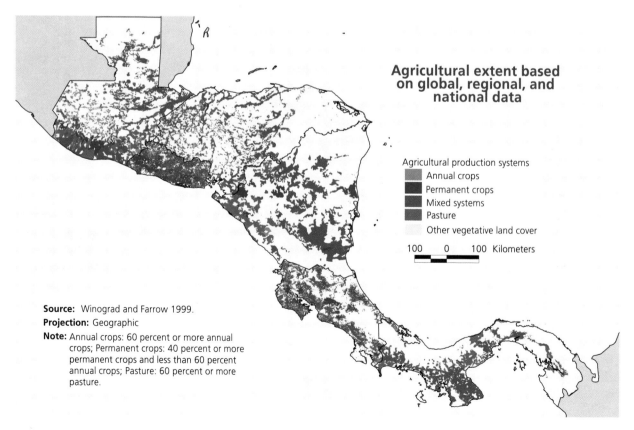

**Agricultural extent based on global, regional, and national data**

Agricultural production systems
- Annual crops
- Permanent crops
- Mixed systems
- Pasture
- Other vegetative land cover

100   0   100   Kilometers

**Source:** Winograd and Farrow 1999.

**Projection:** Geographic

**Note:** Annual crops: 60 percent or more annual crops; Permanent crops: 40 percent or more permanent crops and less than 60 percent annual crops; Pasture: 60 percent or more pasture.

# Map 3a
## Dominant Agroecosystem Types in Central America

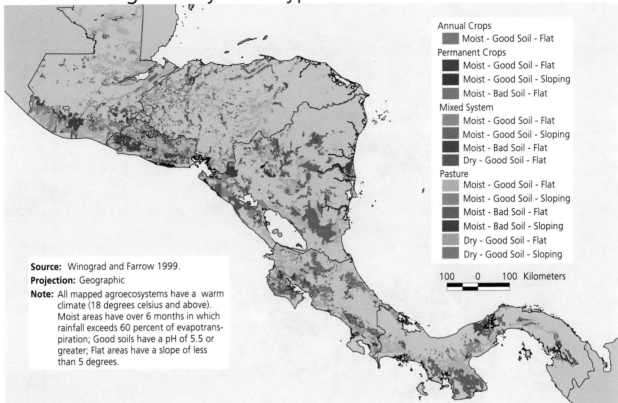

Annual Crops
- Moist - Good Soil - Flat

Permanent Crops
- Moist - Good Soil - Flat
- Moist - Good Soil - Sloping
- Moist - Bad Soil - Flat

Mixed System
- Moist - Good Soil - Flat
- Moist - Good Soil - Sloping
- Moist - Bad Soil - Flat
- Dry - Good Soil - Flat

Pasture
- Moist - Good Soil - Flat
- Moist - Good Soil - Sloping
- Moist - Bad Soil - Flat
- Moist - Bad Soil - Sloping
- Dry - Good Soil - Flat
- Dry - Good Soil - Sloping

100    0    100  Kilometers

**Source:** Winograd and Farrow 1999.

**Projection:** Geographic

**Note:** All mapped agroecosystems have a warm climate (18 degrees celsius and above). Moist areas have over 6 months in which rainfall exceeds 60 percent of evapotranspiration; Good soils have a pH of 5.5 or greater; Flat areas have a slope of less than 5 degrees.

# Map 3b
## Agroecosystem Intensification Levels in Central America

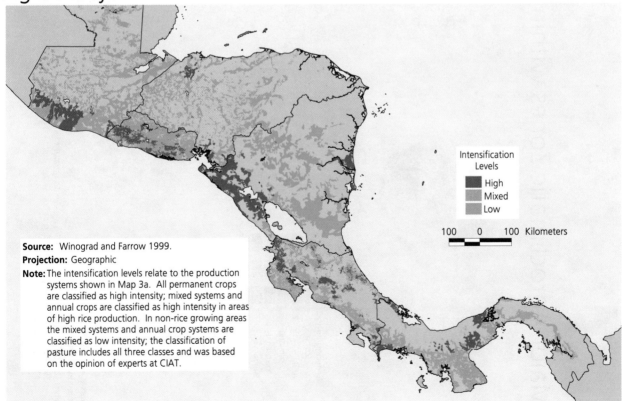

Intensification
Levels
- High
- Mixed
- Low

100    0    100  Kilometers

**Source:** Winograd and Farrow 1999.

**Projection:** Geographic

**Note:** The intensification levels relate to the production systems shown in Map 3a. All permanent crops are classified as high intensity; mixed systems and annual crops are classified as high intensity in areas of high rice production. In non-rice growing areas the mixed systems and annual crop systems are classified as low intensity; the classification of pasture includes all three classes and was based on the opinion of experts at CIAT.

**Map 4**
# Major Agroclimatic Zones within the PAGE Agricultural Extent

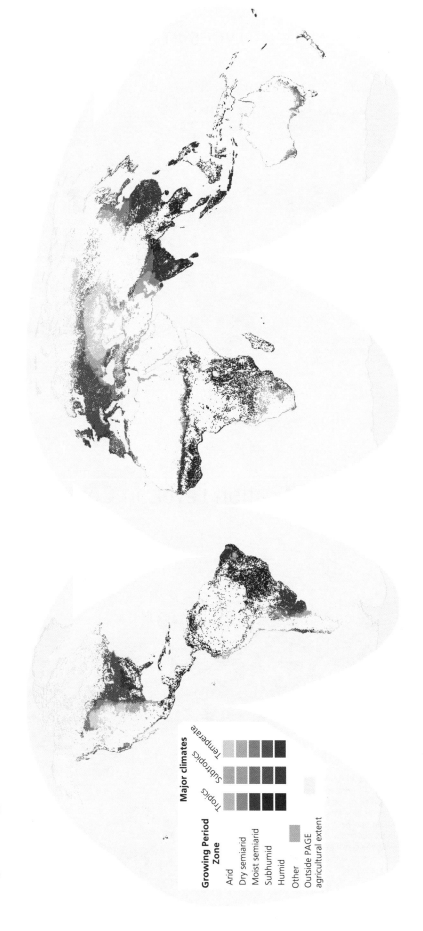

**Major climates**

Tropics  Subtropics  Temperate

**Growing Period Zone**

Arid

Dry semiarid

Moist semiarid

Subhumid

Humid

Other

Outside PAGE agricultural extent

**Source:** FAO/IIASA 1999.

**Projection:** Interrupted Goode's Homolosine

**Notes:** The PAGE agricultural extent includes areas with greater than 30 percent agriculture, based on a reinterpretation of GLCCD 1998 and USGS EDC 1999a, plus additional irrigated areas based on Doell and Siebert 1999. The length of growing period in days for each growing period zone is: Arid, 0 - 74 days; Dry semiarid, 75 - 119 days; Moist semiarid 120 - 179 days; Subhumid, 180 - 269 days; and Humid, 270 - 365 days.

# Map 5
## PAGE Characterization of Global Agroecosystems

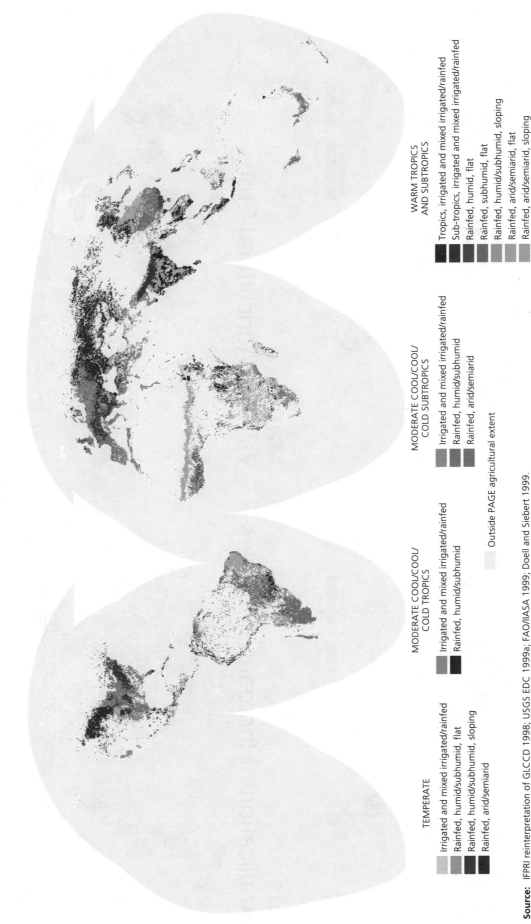

**TEMPERATE**

Irrigated and mixed irrigated/rainfed
Rainfed, humid/subhumid, flat
Rainfed, humid/subhumid, sloping
Rainfed, arid/semiarid

**MODERATE COOL/COOL/ COLD TROPICS**

Irrigated and mixed irrigated/rainfed
Rainfed, humid/subhumid

Outside PAGE agricultural extent

**MODERATE COOL/COOL/ COLD SUBTROPICS**

Irrigated and mixed irrigated/rainfed
Rainfed, humid/subhumid
Rainfed, arid/semiarid

**WARM TROPICS AND SUBTROPICS**

Tropics, irrigated and mixed irrigated/rainfed
Sub-tropics, irrigated and mixed irrigated/rainfed
Rainfed, humid, flat
Rainfed, subhumid, flat
Rainfed, humid/subhumid, sloping
Rainfed, arid/semiarid, flat
Rainfed, arid/semiarid, sloping

**Source:** IFPRI reinterpretation of GLCCD 1998; USGS EDC 1999a; FAO/IIASA 1999; Doell and Siebert 1999.
**Projection:** Interrupted Goode's Homolosine

**Note:** The PAGE agroecosystem characterization includes the PAGE agricultural extent (see Map 1) plus additional irrigated area based on the global map of area equipped for irrigation (see Map 16). Areas with at least 15 percent equipped for irrigation are classified as irrigated. Areas with 5 to 15 percent equipped for irrigation that fall within the PAGE agricultural extent are classified as mixed irrigated/rainfed.

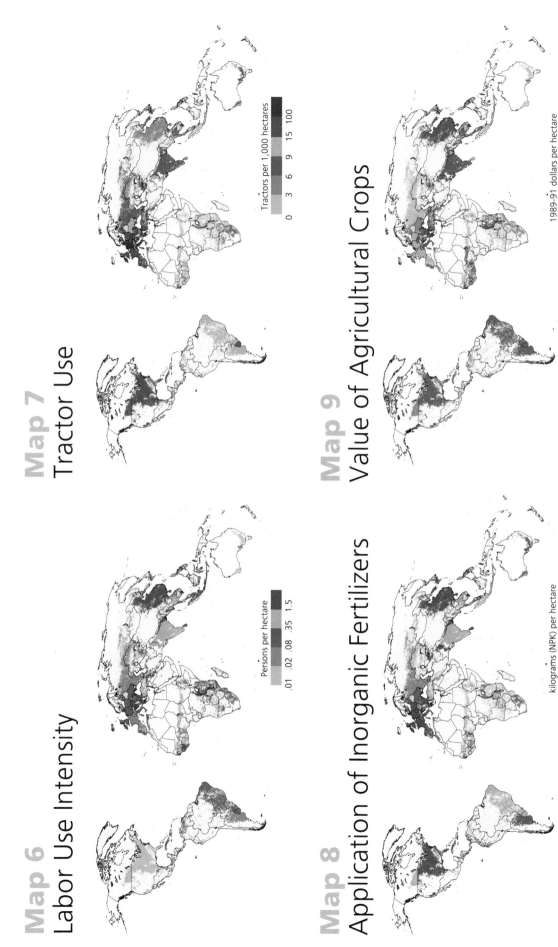

# Map 6
## Labor Use Intensity

Persons per hectare

.01 .02 .08 .35 1.5

# Map 7
## Tractor Use

Tractors per 1,000 hectares

0 3 6 9 15 100

# Map 8
## Application of Inorganic Fertilizers

kilograms (NPK) per hectare

0 10 20 50 100

# Map 9
## Value of Agricultural Crops

1989-91 dollars per hectare

0 200 400 600 1000

Outside PAGE agricultural extent

**Source:** IFPRI calculation based on FAOSTAT 1999.
**Projection:** Interrupted Goode's Homolosine

**Note:** All maps are average national values for 1995–97 shown within the extent of PAGE global agroecosystems (see Map 5). The variables are expressed based on hectares of cropland (annual plus permanent crops). Tractors are defined as all wheel and crawler tractors (excluding garden tractors) used in agriculture. Fertilizers include only commercial inorganic fertilizers: Nitrogen (N), Phosphorus (P2O5) and Potassium (K2O). The value of agricultural crops was calculated by weighting 134 primary crop and 23 primary livestock commodity production quantities by their respective international prices in U.S. dollars for the 1989–91 period.

# Map 10
## Dominant Soil Constraint within the PAGE Agricultural Extent

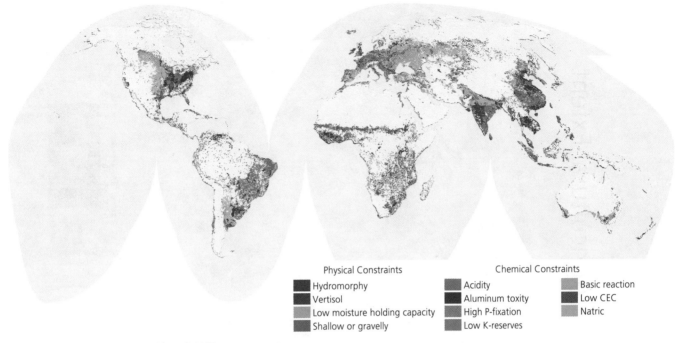

Physical Constraints
- Hydromorphy
- Vertisol
- Low moisture holding capacity
- Shallow or gravelly

Chemical Constraints
- Acidity
- Aluminum toxity
- High P-fixation
- Low K-reserves
- Basic reaction
- Low CEC
- Natric

**Source:** FAO 1995; Smith 1989; Smith et al. 1997.

**Projection:** Interrupted Goode's Homolosine

**Notes:** The PAGE agricultural extent includes areas with greater than 30 percent agriculture, based on a reinterpretation of GLCCD 1998 and USGS EDC 1999a, plus additional irrigated areas based on Doell and Siebert 1999. Dominant soil constraint is the fertility capability classification (FCC) constraint occupying the greatest area proportion of each FAO soil mapping unit.

# Map 11
## Area Free of Soil Constraints within the PAGE Agricultural Extent

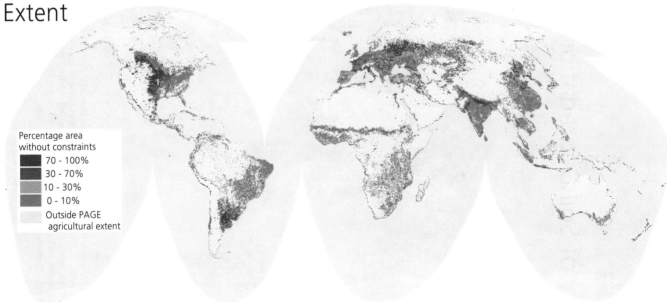

Percentage area
without constraints
- 70 - 100%
- 30 - 70%
- 10 - 30%
- 0 - 10%
- Outside PAGE agricultural extent

**Source:** FAO 1995; Smith 1989; Smith et al. 1997.

**Projection:** Interrupted Goode's Homolosine

**Notes:** The PAGE agricultural extent includes areas with greater than 30 percent agriculture, based on a reinterpretation of GLCCD 1998 and USGS EDC 1999a, plus additional irrigated areas based on Doell and Siebert 1999. The area free of soil constraints is based on the fertility capability classification (FCC) applied to FAO's Digital Soil Map of the World (FAO 1995).

Agroecosystems

## Map 12

# Severity of Soil Degradation within the PAGE Agricultural Extent

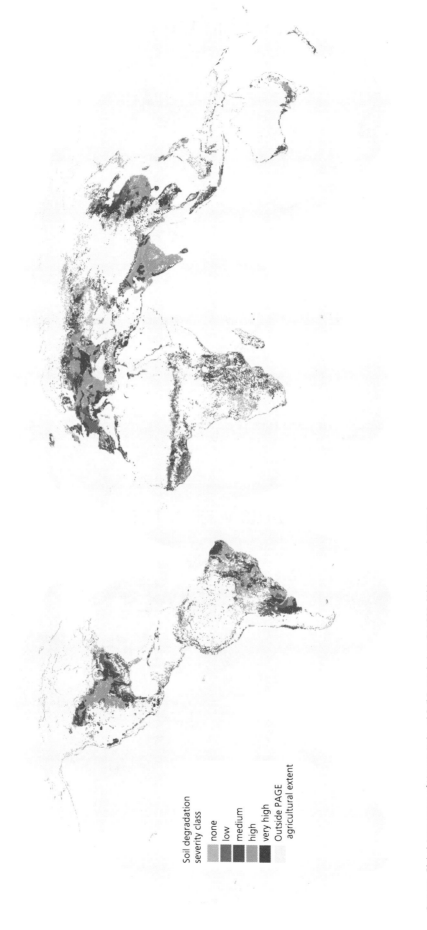

Extent of GLASOD mapping unit affected

0%  5%  10%  25%  50%  100%

light

moderate

strong

extreme

Degree of degradation

Soil degradation
severity class

none
low
medium
high
very high
Outside PAGE
agricultural extent

**Source:** Global Assessment of Human Induced Soil Degradation: GLASOD (Oldeman et al. 1991a).

**Projection:** Interrupted Goode's Homolosine

**Note:** The PAGE agricultural extent includes areas with greater than 30 percent agriculture, based on a reinterpretation of GLCCD 1998 and USGS EDC 1999a, plus additional irrigated areas based on Doell and Siebert 1999. The actual amount to which the degraded areas within GLASOD mapping units physically overlap agricultural areas within the PAGE agricultural extent is unknown (see section on Soil Quality Status and Change). As shown, the soil degradation severity class is a composite based on the degree of degradation and the extent of GLASOD mapping unit affected.

# Map 13
## Soil Degradation within the PAGE Agricultural Extent of South and Southeast Asia

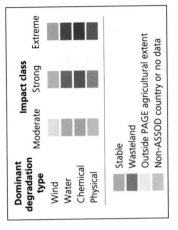

**Dominant degradation type**

Impact class: Moderate, Strong, Extreme

- Wind
- Water
- Chemical
- Physical

- Stable
- Wasteland
- Outside PAGE agricultural extent
- Non-ASSOD country or no data

**Source:** Assessment of the Status of Human-Induced Soil Degradation in South and Southeast Asia (ASSOD) (van Lynden and Oldeman 1997).

**Projection:** Geographic

**Note:** The PAGE agricultural extent includes areas with greater than 30 percent agriculture, based on a reinterpretation of GLCCD 1998, plus additional irrigated areas based on Doell and Siebert 1999. Wind erosion includes: loss of topsoil by wind action, terrain deformation, and overblowing; water erosion includes: loss of topsoil and terrain deformation; chemical degradation includes: fertility decline and reduced organic matter content, salinization/alkalinization, dystrification/acidification, eutrophication, and pollution; physical degradation includes: compaction, crusting and sealing, waterlogging, lowering of the soil's surface, loss of productive function, and aridification; stable lands include land stable under natural conditions or under human influence with no or little degradation impact; and wasteland is land without vegetation and with the near absense of human influence on the soil stability.

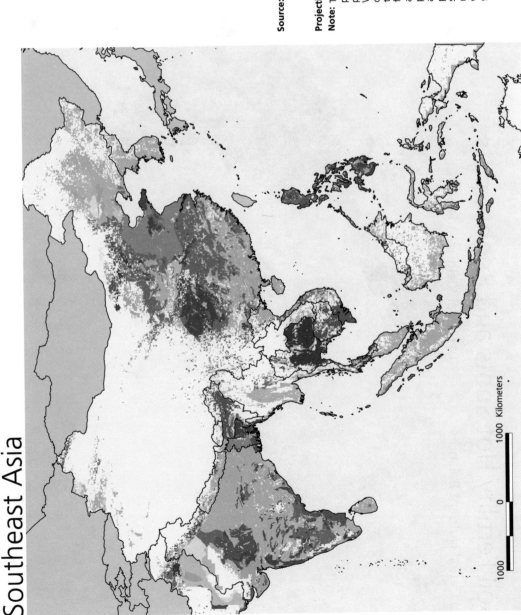

0    1000    Kilometers

## Map 14
# Soil Fertility Hot Spots and Bright Spots

(a) Cereal Nutrient Balances

(b) Cereal Yield Trends

(c) Potential Hot Spots and Bright Spots

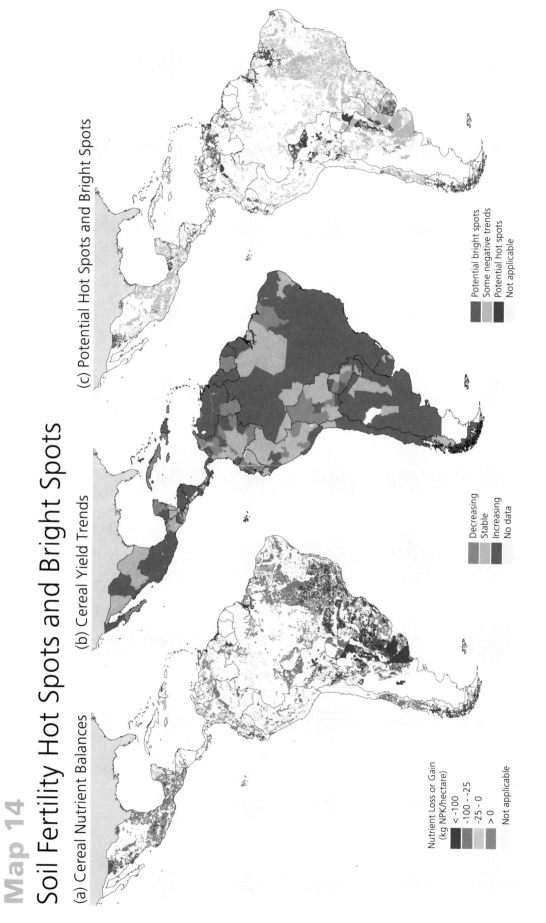

Nutrient Loss or Gain
(kg NPK/hectare)
- < -100
- -100 - -25
- -25 - 0
- > 0
- Not applicable

Decreasing
Stable
Increasing
No data

Potential bright spots
Some negative trends
Potential hot spots
Not applicable

**Source:** Sebastian and Wood 2000.
**Projection:** Geographic
**Note:** Cereal nutrient balances are estimated as the difference between mineral and organic fertilizer application and crop residue recycling for cereals (inputs) and the nutrients extracted in cereal grain (outputs). Nutrient balances were allocated to specific geographic areas using subnational 1993-95 production statistics and information on climate, soil, and elevation. Cereal yield trends are based on subnational 1975-95 data for rice, wheat, maize, and sorghum. The map of potential hot spots and bright spots combines the map of nutrient balances and the map of cereal yield trends.

# Map 15
## Variability in Length of Growing Period within the PAGE Agricultural Extent

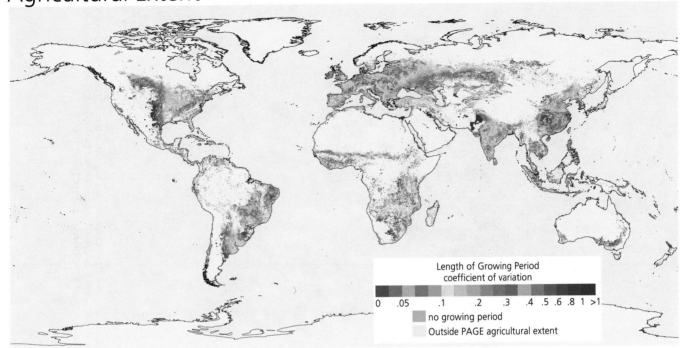

**Source:** FAO/IIASA 1999.
**Projection:** Geographic
**Note:** The PAGE agricultural extent includes areas with greater than 30 percent agriculture, based on a reinterpretation of GLCCD 1998 and USGS EDC 1999a, plus additional irrigated areas based on Doell and Siebert 1999. The length of growing period (LGP) is the number of days per year in which moisture and temperature conditions will support plant growth. The year-to-year variability is calculated for the 30 year period, 1960-90.

# Map 16
## Area Equipped for Irrigation by Growing Period Zone

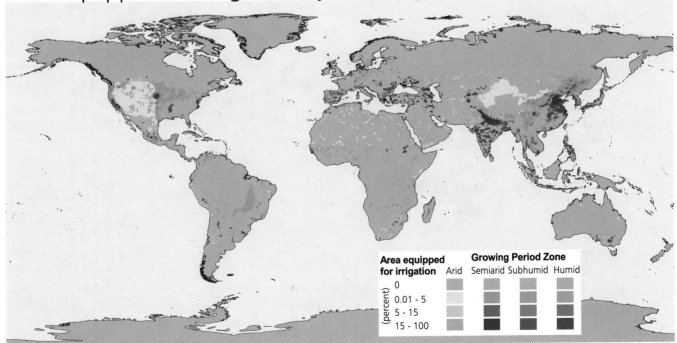

**Source:** Doell and Siebert 1999; FAO/IIASA 1999.
**Projection:** Geographic
**Note:** The percent area equipped for irrigation is calculated within a 50 by 50 kilometer area. The length of growing period in days for each growing period zone is: Arid, 0 - 74 days; Semiarid, 75 - 179 days; Subhumid, 180 - 269 days; and Humid, 270 - 365 days.

## Map 17

# Agricultural Share of Watershed Area

Agricultural share
of total basin area

0 - 10%
10 - 20%
20 - 40%
40 - 60%
> 60%

**Source:** UNH - GRDC Runoff Database (UNH - GRDC 1999).

**Projection:** Interrupted Goode's Homolosine

**Note:** The share of each watershed that is agricultural was calculated by applying a weighted percentage to each PAGE agricultural land cover class (80 percent for areas with at least 60 percent agriculture; 50 percent for areas with 40-60 percent agriculture; 35 percent for areas with 30-40 percent agriculture; and 5 percent for areas with 0-30 percent agriculture) to determine the total agricultural area within a watershed. Only watersheds with 10 percent or greater agricultural share are mapped. The agricultural shares do not include additional irrigated areas based on Doell and Siebert 1999.

# Map 18
## Agricultural Share of Protected Areas

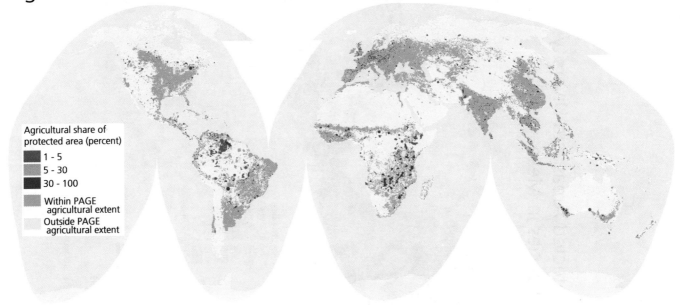

Agricultural share of
protected area (percent)
- 1 - 5
- 5 - 30
- 30 - 100
- Within PAGE
  agricultural extent
- Outside PAGE
  agricultural extent

**Source:** WCMC 1999.

**Projection:** Interrupted Goode's Homolosine

**Notes:** The PAGE agricultural extent includes areas with greater than 30 percent agriculture, based on a reinterpretation of GLCCD 1998 and USGS EDC 1999a, plus additional irrigated areas based on Doell and Siebert 1999. The share of protected areas that is agricultural was calculated for each protected area using the reinterpreted EDC Seasonal Land Cover Characterization database. For protected areas represented only by points, a circular buffer was generated corresponding to the size of the protected area.

# Map 19
## Percentage Tree Cover within the PAGE Agricultural Extent

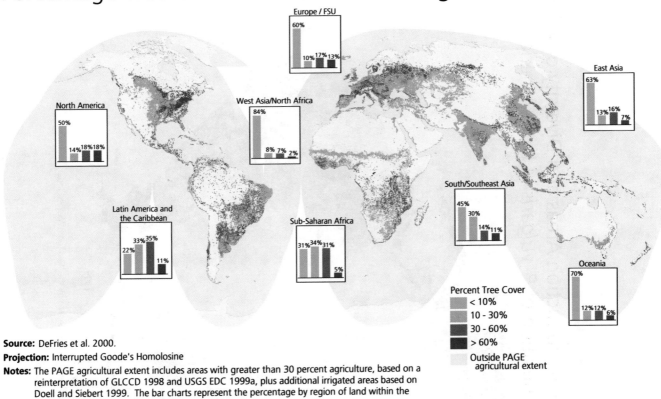

Percent Tree Cover
- < 10%
- 10 - 30%
- 30 - 60%
- > 60%
- Outside PAGE
  agricultural extent

**Source:** DeFries et al. 2000.

**Projection:** Interrupted Goode's Homolosine

**Notes:** The PAGE agricultural extent includes areas with greater than 30 percent agriculture, based on a reinterpretation of GLCCD 1998 and USGS EDC 1999a, plus additional irrigated areas based on Doell and Siebert 1999. The bar charts represent the percentage by region of land within the PAGE agricultural extent for each tree cover class.

Agroecosystems

## Map 21

### Urban and Protected Areas and Diversity of *Phaseolus* Species in Mesoamerica

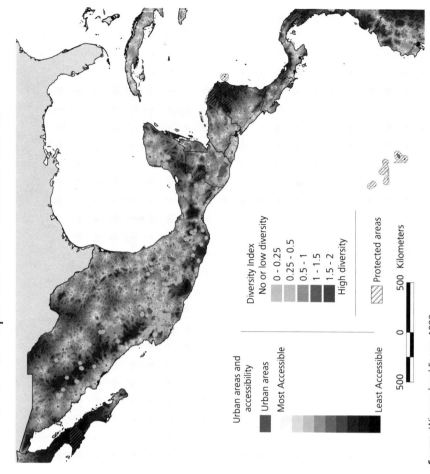

Urban areas and accessibility

Urban areas

Most Accessible

Least Accessible

Diversity Index
No or low diversity

0 - 0.25
0.25 - 0.5
0.5 - 1
1 - 1.5
1.5 - 2
High diversity

Protected areas

500   0   500   Kilometers

**Source:** Winograd and Farrow 1999.
**Projection:** Geographic
**Note:** Diversity is represented by the 'Shannon-Weaver' index.

## Map 20

### Bean Production Areas and Diversity of Wild Population of *P. vulgaris* in Latin America

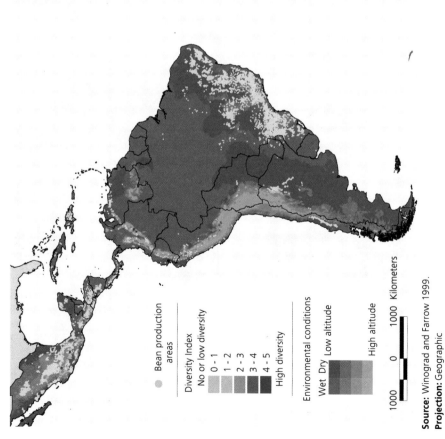

Bean production areas

Diversity Index
No or low diversity

0 - 1
1 - 2
2 - 3
3 - 4
4 - 5
High diversity

Environmental conditions

Wet   Dry   Low altitude

High altitude

1000   0   1000   Kilometers

**Source:** Winograd and Farrow 1999.
**Projection:** Geographic
**Note:** Diversity is represented by the 'Shannon-Weaver' index.

# Map 22

## Carbon Storage in Above- and Below-Ground Live Vegetation within the PAGE Agricultural Extent

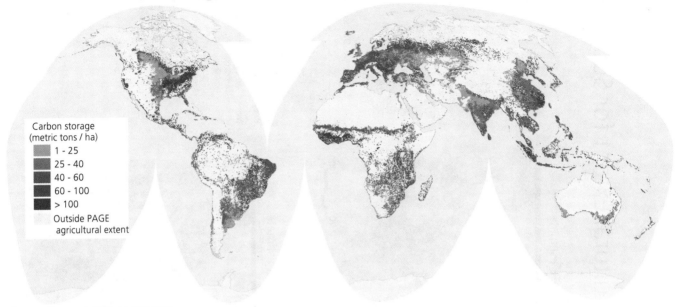

Carbon storage
(metric tons / ha)
- 1 - 25
- 25 - 40
- 40 - 60
- 60 - 100
- > 100
- Outside PAGE agricultural extent

Source: Olson et al. 1983; GLCCD 1998.

Projection: Interrupted Goode's Homolosine

Notes: The PAGE agricultural extent includes areas with greater than 30 percent agriculture, based on a reinterpretation of GLCCD 1998 and USGS EDC 1999a, plus additional irrigated areas based on Doell and Siebert 1999. Carbon stored in above- and below-ground live vegetation is based on estimates developed by Olson et al. (1983) for the dominant vegetation types found in the world's major ecosystems and applied to the Global Land Cover Characteristics Database (GLCCD) at a resolution of 1km by 1km.

# Map 23

## Carbon Storage in Soils within the PAGE Agricultural Extent

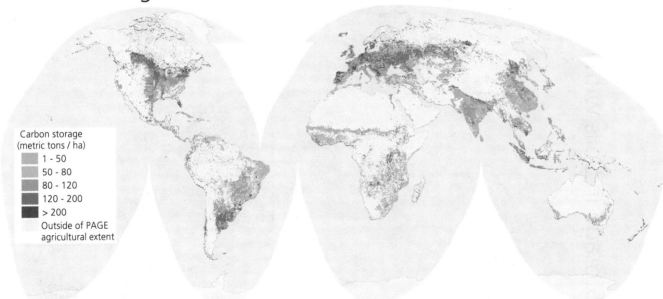

Carbon storage
(metric tons / ha)
- 1 - 50
- 50 - 80
- 80 - 120
- 120 - 200
- > 200
- Outside of PAGE agricultural extent

Source: IFPRI calculation based on: Batjes 1996; Batjes 2000; FAO 1995; GLCCD 1998 and USGS EDC 1999a.

Projection: Interrupted Goode's Homolosine

Notes: The PAGE agricultural extent includes areas with greater than 30 percent agriculture, based on a reinterpretation of GLCCD 1998 and USGS EDC 1999a, plus additional irrigated areas based on Doell and Siebert 1999. Batjes estimated the average soil organic carbon (SOC) content at a depth of 100cm by soil type for over 4,000 individual soil profiles contained in the World Inventory of Soil Emission Potentials (WISE) database (Batjes 1996; Batjes and Bridges 1994). The authors calculated the global estimate of SOC storage by applying Batjes' (1996 and 2000) SOC content values by soil type area share of each 5 x 5 minute unit of the Digital Soil Map of the World (FAO 1995).

Agroecosystems

# Map 24

# Carbon Storage in Soils within the PAGE Agricultural Extent for South America

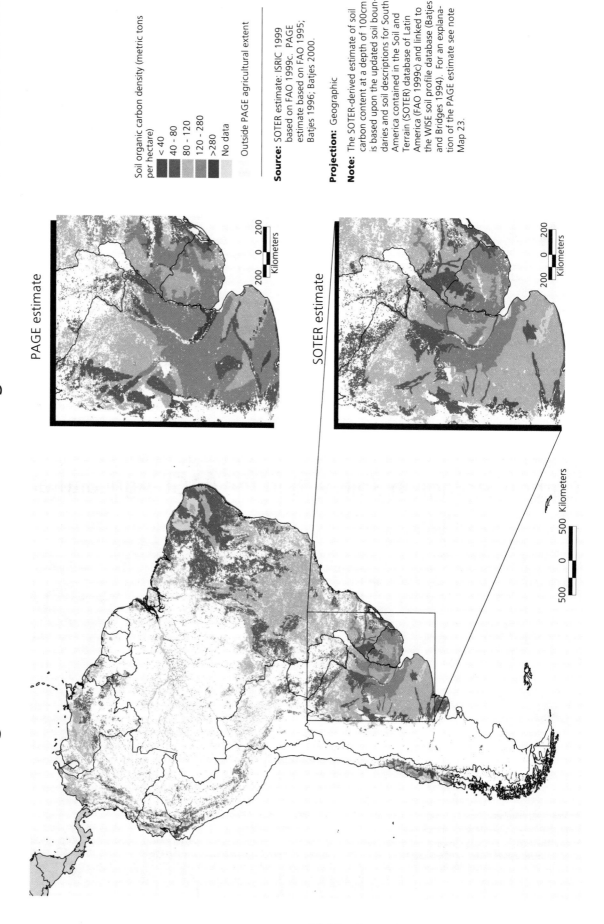

Soil organic carbon density (metric tons per hectare)

- < 40
- 40 - 80
- 80 - 120
- 120 - 280
- >280
- No data

Outside PAGE agricultural extent

PAGE estimate

SOTER estimate

**Source:** SOTER estimate: ISRIC 1999 based on FAO 1999c. PAGE estimate based on FAO 1995; Batjes 1996; Batjes 2000.

**Projection:** Geographic

**Note:** The SOTER-derived estimate of soil carbon content at a depth of 100cm is based upon the updated soil boundaries and soil descriptions for South America contained in the Soil and Terrain (SOTER) database of Latin America (FAO 1999c) and linked to the WISE soil profile database (Batjes and Bridges 1994). For an explanation of the PAGE estimate see note Map 23.